FLAVOR, FRAGRANCE, and ODOR ANALYSIS

second edition

FLAVOR, FRAGRANCE, and ODOR ANALYSIS

second edition

edited by
Ray Marsili

CRC Press
Taylor & Francis Group
Boca Raton   London   New York

CRC Press is an imprint of the
Taylor & Francis Group, an **informa** business

CRC Press
Taylor & Francis Group
6000 Broken Sound Parkway NW, Suite 300
Boca Raton, FL 33487-2742

First issued in paperback 2016

Version Date: 20110810

ISBN 13: 978-1-138-19857-9 (pbk)
ISBN 13: 978-1-4398-4673-5 (hbk)

Library of Congress Cataloging-in-Publication Data

Flavor, fragrance, and odor analysis / editor, Ray Marsili. -- 2nd ed.
  p. cm.
  Includes bibliographical references and index.
  ISBN 978-1-4398-4673-5 (hardback)
  1. Food--Sensory evaluation. I. Marsili, Ray, 1946- II. Title.

TX546.F52 2011
664'.07--dc23   2011031500

Visit the Taylor & Francis Web site at
http://www.taylorandfrancis.com

and the CRC Press Web site at
http://www.crcpress.com

Dedication

To my newest grandchild: Nyla

Contents

Preface

Sample preparation techniques for isolating and concentrating flavor-active and odor-active chemicals from foods prior to performing gas chromatography–mass spectrometry (GC–MS) analysis continue to evolve, providing lower detection limits while becoming more amenable to instrumental automation. Solventless extraction techniques like solid-phase microextraction (SPME) and stir bar sorptive extraction (SBSE) have wide application appeal and are now significant problem-solving tools for flavor and odor chemists. Whereas the first edition of *Flavor, Fragrance, and Odor Analysis* emphasized SPME, a new technique at the time, this edition focuses on the advantages of SBSE, including sequential SBSE and other sample manipulation techniques that increase the sensitivity and application potential of this surprisingly simple but amazingly powerful technique. Applications are emphasized. It should not go unmentioned that instrumental offerings by Gerstel, Inc., Mülheim an der Ruhr, Germany, which include automated thermal desorption units, the Cooled Inlet System (CIS), and the MultiPurpose Sampler (MPS), have made SPME and SBSE automation a practical reality.

Chapter 2 discusses a simplified method for switching from one-dimensional to two-dimensional GC–MS. Chapter 5 discusses a technique for measuring synergy effects between odorants. In some respects most GC–olfactometry studies conducted in the past have been oversimplified, because they do not consider possible synergy effects between odorants. This has been largely neglected by researchers, but it should be a highly significant field of GC–olfactometry research in the future.

The final chapter, Chapter 9, on character-impact compounds is an expanded, updated version of a similar chapter that appeared in the first edition of *Flavor, Fragrance, and Odor Analysis*. Many researchers have indicated that this chapter, placed at the end of the book for quick reference, is the best treatment of the topic that they have seen. This chapter, which was previously published in *Sensory-Directed Flavor Analysis*, has been updated by the author.

I thank the contributing authors for their excellent chapters. Their chapter contributions contain valuable information for flavor and odor researchers. Also, I thank my wife, Deborah, for her enormous support and acknowledge the laboratory support offered to me by my eldest grandson Charles R. Laskonis, who has worked with me for the past several years and has great potential as a flavor chemist.

Contributors

William M. Coleman, III
iii Consulting, LLC
Winston-Salem, North Carolina

Amanda Kowalsick
Department of Chemistry
Tufts University
Medford, Massachusetts

Patricio R. Lozano
Kerry Ingredients & Flavours
Americas Region
Beloit, Wisconsin

Ray Marsili
Marsili Consulting Group
Rockford College
Rockford, Illinois

Robert J. McGorrin
Food Science and Technology
Oregon State University
Corvallis, Oregon

Yvette Naudé
Chromatography–Mass
    Spectrometry Group
Department of Chemistry
University of Pretoria
Pretoria, South Africa

Nobuo Ochiai
Application Development
Gerstel K. K.
Tokyo, Japan

Albert Robbat, Jr.
Department of Chemistry
Tufts University
Medford, Massachusetts

Egmont R. Rohwer
Chromatography–Mass
    Spectrometry Group
Department of Chemistry
University of Pretoria
Pretoria, South Africa

chapter one

Sequential stir bar sorptive extraction

Nobuo Ochiai

Contents

Introduction

In odor analysis, gas chromatography (GC) techniques are used as the main analytical method because the majority of odor compounds are volatiles. In order to determine odor compounds in various sample matrices at trace levels, a GC method usually requires sample preparation (e.g., an extraction and enrichment step) before GC analysis. However, no single sample preparation technique is appropriate for every type of analyte or matrix. In the past decades, the development of miniaturized and solvent-less (or solvent-minimized) sample preparation methods has been recognized as one of the most important projects because it provides many benefits such as high sensitivity, small sample volume, fast analysis, high sample throughput, low operational cost, and low solvent consumption. As a typical example of the miniaturized sample preparation methods, solid-phase microextraction (SPME) (Arthur and Pawliszyn 1990) and stir bar sorptive extraction (SBSE) (Baltussen et al. 1999) have been developed in the last decades. These methods are simple, solventless techniques allowing the extraction and concentration in a single step. Also, these methods provide enhanced sensitivity because the extracted fraction (on a fiber or on a stir bar) can be introduced quantitatively into a GC system by thermal desorption (TD). Although both SPME and SBSE can be used in immersion mode and in headspace mode, for odor analysis, SPME is mainly used in the headspace mode (HS-SPME), whereas SBSE is most often used in the immersion mode. It is known that the enrichment factor for SBSE, which is determined by the analyte recovery in the extraction phase (polydimethylsiloxane [PDMS]), is higher than that of SPME because of 50–250 times larger volume of extraction phase on the stir bar. SBSE has been successfully applied to analysis of odor compounds in various sample matrices, for example, water, beverages, fruits, herbs, plant material, essential oils, and vinegar (David and Sandra 2007).

This chapter discusses a novel extraction procedure for SBSE termed sequential SBSE. It will attempt a comparison between conventional SBSE and sequential SBSE for the analysis of odor compounds in aqueous samples. The examples show more uniform enrichment of odor compounds covering a wide polarity range. Also, sequential SBSE with a new phase stir bar, which has a polyethylene-glycol (PEG)-modified silicon coating, and a conventional PDMS stir bar will be demonstrated.

Sequential stir bar sorptive extraction

SBSE recovery with a PDMS stir bar for water sample can be estimated if the octanol–water distribution coefficient (K_{ow}) of the analyte is known. The K_{ow} is the ratio of the concentration of a chemical in octanol and in water at a specified temperature. Because the K_{ow} is proposed to the PDMS–water distribution coefficient (David and Sandra 2007), the K_{ow} gives a good indication of the applicability of SBSE for a given analyte in aqueous samples. The mass of analyte extracted into the PDMS phase at full equilibration (expected recovery) is calculated by the following equations.

$$K_{ow} \approx K_{PDMSw} = C_{PDMS}/C_w = (m_{PDMS}/m_w)(V_w/V_{PDMS})$$

where C_{PDMS} is the analyte concentration in PDMS, C_w is the analyte concentration in water, m_{PDMS} is the mass of analyte in PDMS, m_w is the mass of analyte in water, V_{PDMS} is the volume of PDMS, and V_w is the volume of water. $V_w/V_{PDMS} = \beta$, the phase ratio of the water–PDMS system:

$$K_{ow}/\beta = m_{PDMS}/m_w = m_{PDMS}/(m_0 - m_{PDMS})$$

$$\text{Recovery} = m_{PDMS}/m_0 = K_{ow}/\beta/(1 + K_{ow}/\beta)$$

where m_0 is the total amount of analyte originally present in the water sample.

Hydrophobic solutes with a high K_{ow} can be extracted with high recovery, whereas hydrophilic solutes with a low K_{ow}, for example, polar compounds, show lower recovery (Baltussen et al. 1999). In order to increase recovery of more hydrophilic solutes, one could use salt addition, for example, 20%–30% NaCl. However, salt addition resulted in decreasing recovery of more hydrophobic solutes (Leon et al. 2003; Nakamura and Daishima 2005). Salt addition in SBSE using a single stir bar will therefore have limited benefit when developing a method for simultaneous analysis that includes compounds of widely varying polarities. Therefore, with conventional SBSE using a single-step extraction, one has to find a compromise of extraction condition. Ochiai et al. (2008) developed an SBSE procedure termed sequential SBSE for more uniform enrichment of organic pollutants in aqueous samples, which is performed sequentially for one aliquot under two extraction conditions using two stir bars. The first extraction with unmodified sample is mainly targeting solutes with high K_{ow} (log K_{ow} > 4.0), the second extraction with modified sample solution (containing 30% NaCl) is targeting solutes with low and medium K_{ow} (log K_{ow} < 4.0). After extraction, the two stir bars are placed in a single glass desorption liner and are simultaneously desorbed. This technique counters the difficulty in recovering both hydrophobic and hydrophilic compounds. This approach produced a remarkable improvement in recovery for a set of 80 pesticides (log K_{ow}: 1.70–8.35) in water, 82%–113% for the majority and less than 80% for only five hydrophilic compounds.

Marsili et al. (2009) applied sequential SBSE to the analysis of odor compounds, which contribute to the "wet-dog" malodor, in casein powder. Twenty-five milliliters of aqueous solution which contains 1 g of casein was extracted with sequential SBSE using buffer and salt as the modifiers. More than 60 odor compounds including off-flavor compounds (e.g., trihaloanisoles) were detected. Quantification was performed for 31 odor compounds (e.g., trihaloanisoles, geosmin, methional, guaiacol, indole, skatole, E-2-nonenal) using a standard addition calibration method. The method showed very good linearity in the range of 20–6000 ng/g (r^2 > 0.9905) and repeatability (relative standard deviation [RSD] < 6.5%).

Instrumentation

The TD-GC–MS analysis was performed with a thermal-desorption unit (TDU) equipped with an MPS 2 autosampler and a CIS 4 programmed temperature vaporization (PTV) inlet (Gerstel, Mülheim an der Ruhr, Germany) installed on an Agilent 7890A gas chromatograph with a 5975 mass-selective detector (Agilent Technologies, California). A nitrogen phosphorus detector (NPD, Agilent) was also installed on the GC–MS system.

Sequential stir bar sorptive extraction

Stir bars coated with 24 μL of PDMS (Twister) and 32 μL of PEG Silicon (EG Silicon Twister) were obtained from Gerstel. For the first SBSE, 5 mL of aqueous sample were transferred to 10-mL headspace vials. A stir bar was added, and the vial was sealed with a screw cap. SBSE of several samples was performed simultaneously at room temperature (24°C) for 60 min while stirring at 1500 rpm with a multiple position magnetic stirrer (20 positions) from Global Change (Tokyo, Japan). After the first extraction, the stir bar was removed with forceps, dipped briefly in Milli-Q water, dried with a lint-free tissue, and placed in a glass TD liner. The glass liner was temporarily placed and stored in a sealed sample tray of the MPS2. For the second extraction, 30% NaCl was dissolved in the sample. Then, a second stir bar was added, and the vial was capped again. The second extraction was performed under the same conditions as the first extraction. After the second extraction, the stir bar was removed with forceps, dipped briefly in Milli-Q water, dried with a lint-free tissue, and placed in the glass liner which contained the first SBSE stir bar. Finally, the glass liner was placed in the TD unit. No further sample preparation was necessary. Conventional SBSE with or without the modifier (e.g., 30% NaCl or methanol) was performed for 2 h as a comparison. Figure 1.1 shows the sequential SBSE procedure.

Sequential SBSE using a single stir bar is also possible when using 30% NaCl addition as the modifier. However, if the second extraction is performed with different kinds of modifiers such as organic solvent (e.g., 20% methanol), derivatizing reagent, or pH adjustment, then using two different stir bars is preferred because the modifiers may influence the solutes absorbed in the stir bar with the first extraction. Also, sequential SBSE using two stir bars can be performed with two different types of stir bars, for example, a conventional PDMS stir bar and a new phase stir bar, which has a more polar EG Silicon coating.

Reconditioning of stir bars was done after use by soaking in Milli-Q purified water and a mixture of methylene chloride–methanol (1:1) for 24 h each; stir bars were then removed from the solvent and dried on a clean surface at room temperature for 1 h. Finally, the stir bars were thermally conditioned for 10 min at 250°C (PDMS stir bar) or at 220°C (EG Silicon stir bar) in a flow of helium.

TD-GC–(NPD)/MS

The two stir bars were thermally desorbed by programming the TDU from 40°C (held for 1 min) to 220°C (held for 5 min) at 720°C/min with 50 mL/min desorption flow. Solvent vent mode of the TDU was initially performed for 1 min at 40°C to remove residual water from the EG Silicon

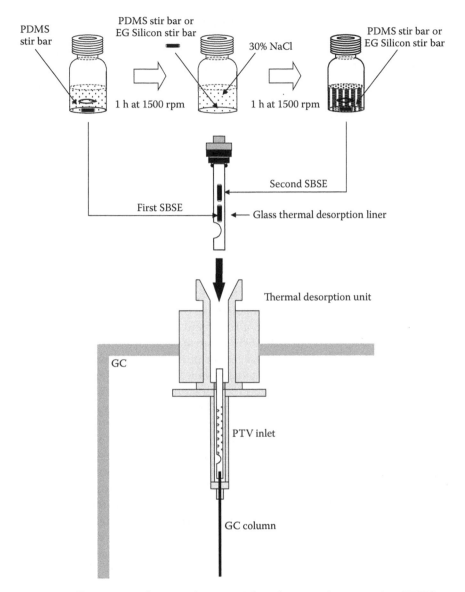

Figure 1.1 Experimental setup of sequential stir bar sorptive extraction (SBSE).

stir bar. Desorbed compounds were focused at 20°C on a Tenax TA packed liner in the PTV inlet for subsequent GC–MS analysis. After desorption, the PTV inlet was programmed from 20°C to 240°C (held for 5 min) at 720°C/min to inject trapped compounds onto the analytical column. The injection was performed in the split mode with a split ratio of 1:15. The separation was performed with helium carrier gas on a DB-Wax fused silica capillary column (30 m × 0.25 mm i.d., 0.25-μm film thickness, Agilent Technologies). The oven temperature was programmed from 40°C (held for 2 min) at 10°C/min to 240°C (held for 10 min). The mass spectrometer was operated in scan mode using electron ionization (electron-accelerating voltage: 70 V). Scan range was set from m/z 29 to 300 and sampling rate of two, resulting in a scan rate of 2.68 scans per second. For simultaneous NPD and MS detection, the NPD temperature was set at 325°C, and its flow rate was 3.0, 5.0, and 60.0 mL/min for hydrogen, helium makeup, and air, respectively. The split ratio to the MS and NPD was set to 1:1.

Extraction efficiency of odor compounds

In order to evaluate extraction efficiency of sequential SBSE for odor compounds in aqueous samples, the recovery of sequential SBSE for 16 model compounds including various types of odor compounds (e.g., alcohol, aldehyde, ester, lactone, monoterpene, monoterpenoid, and phenol) in water were compared with those of conventional SBSE with or without salt addition. The log K_{ow} values of the model compounds, which were calculated with an SRC-KOWWIN software package (Syracuse Research, Syracuse, New York) according to a fragment constant estimation methodology (Meylan and Haward 1995), were in the range of 1.34 (guaiacol) to 4.83 (limonene). The concentration of the model compounds was 100 ng/mL each. The recovery was calculated by comparing peak areas with those of a direct liquid injection of a standard solution for calibration curves, which was injected into the TDU system (through a septum head) containing two stir bars in a glass desorption liner. Figure 1.2 shows a comparison of the total ion chromatogram (TIC) of a spiked sample obtained by (a) conventional SBSE without modifier, (b) conventional SBSE with salt addition (30% NaCl), and (c) sequential SBSE. Table 1.1 shows theoretical SBSE recovery with a 5-mL sample volume and 24 μL of PDMS volume, SBSE recovery without modifier, SBSE recovery with 30% NaCl, and sequential SBSE recovery. For conventional SBSE without salt addition, although the compounds with log K_{ow} of more than 2.8 (e.g., ethyl hexanoate, nonanal, ethyl octanoate, β-damascenone, and limonene) except alcohols show good correspondence with the theoretical recovery, large deviations are observed for the rest of the compounds (specifically for alcohol and phenol, which have hydroxyl function). For conventional SBSE with salt addition, the recovery for compounds with log K_{ow} of less than 3.6 dramatically increased with salt addition, for example, for hexanal (log K_{ow}: 1.80), γ-nonalactone (log K_{ow}: 2.08), and linalool (log K_{ow}: 3.38), the recovery increased from 15% to 43%, 28% to 62%, and 41% to 80%, respectively. However, recovery for compounds with log K_{ow} of more than 3.8 drastically decreased, for example, for ethyl octanoate (log K_{ow}: 3.81), and limonene (log K_{ow}: 4.83), the recovery decreased

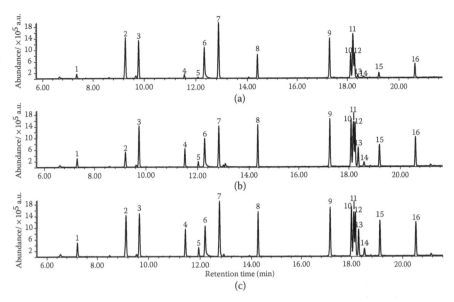

Figure 1.2 Total ion chromatograms of spiked water samples obtained by (a) conventional SBSE without modifier, (b) conventional SBSE with salt addition (30% NaCl), and (c) sequential SBSE. 1. Hexanal (log K_{ow}: 1.80); 2. Limonene (log K_{ow}: 4.83); 3. Ethyl hexanoate (log K_{ow}: 2.83); 4. Hexanol (log K_{ow}: 1.82); 5. 3-Hexenol (log K_{ow}: 1.61); 6. Nonanal (log K_{ow}: 3.27); 7. Ethyl octanoate (log K_{ow}: 3.81); 8. Linalool (log K_{ow}: 3.38); 9. Citronellol (log K_{ow}: 3.56); 10. Phenethyl acetate (log K_{ow}: 2.57); 11. β-Damascenone (log K_{ow}: 4.21); 12. Geraniol (log K_{ow}: 3.47); 13. p-Cymen-8-ol (log K_{ow}: 2.49); 14. Guaiacol (log K_{ow}: 1.34); 15. Phenethyl alcohol (log K_{ow}: 1.57); 16. γ-Nonalactone.

from 100% to 67%, and 94% to 28%, respectively. In contrast with conventional SBSE with or without salt addition, the sequential approach could eliminate the negative effect of the salt for compounds with log K_{ow} of more than 3.8, while maintaining increased recovery for hydrophilic solutes with salt addition, resulting in high recovery in the range of 90%–100% for nine compounds with log K_{ow} of more than 2.5. Although the recovery for the seven compounds with log K_{ow} of less than 2.5 was in the range of 7.6%–85%, these seven compounds show good correspondence with the original theoretical recovery or even higher recovery (e.g., hexanal and γ-nonalactone).

Figure 1.3 shows a comparison of the TIC of the single malt whiskey "B 12 years" (fivefold diluted with natural water) obtained by (a) conventional SBSE without modifier, (b) conventional SBSE with salt addition (30% NaCl), and (c) sequential SBSE. It is interesting to observe that several hydrophobic compounds with log K_{ow} of more than 4.0, for example, ethyl nonanoate (log K_{ow}: 4.30), isoamyl octanoate

Table 1.1 Comparison of SBSE Recovery (%) for the Model Odor
Compounds in Water at 100 ng/mL

Compound	log K_{ow}[b]	Recovery (%)			
		Theoretical[c]	PDMS[d]	PDMS-salt[e]	Sequential SBSE[f]
Guaiacol	1.34	9.5	1.5	5.6	9.3
Phenethyl alcohol[a]	1.57	15	0.9	4.2	7.6
3-Hexenol	1.61	16	1.2	6.7	12
Hexanal	1.80	23	15	43	59
Hexanol	1.82	24	3.3	19	32
γ-Nonalactone	2.08	37	28	62	85
p-Cymen-8-ol	2.49	60	4.6	34	49
Phenethyl acetate	2.57	64	40	85	99
Ethyl hexanoate	2.83	76	77	86	97
Nonanal	3.27	90	89	80	94
Linalool	3.38	92	41	80	94
Geraniol	3.47	93	42	82	94
Citronellol	3.56	95	73	89	101
Ethyl octanoate	3.81	97	103	67	103
β-Damascenone	4.21	99	102	85	90
Limonene	4.83	100	94	28	96

PDMS = polydimethylsiloxane; SBSE = stir bar sorptive extraction.

[a] Concentration: 1000 ng/mL.
[b] The log K_{ow} values are calculated with SRC-KOWWIN software.
[c] Theoretical SBSE recovery with 5-mL sample and 24 μL PDMS.
[d] SBSE without modifier.
[e] SBSE with salt addition (30% NaCl).
[f] Sequential SBSE consisting of the first extraction without modifier and the second extraction with salt addition (30% NaCl).

(log K_{ow}: 5.24), ethyl dodecanoate (log K_{ow}: 5.78), nerolidol (log K_{ow}: 5.68), and farnesol (log K_{ow}: 5.77), detected in the chromatogram (a) at retention time (RT) of 14.18, 15.99, 18.23, 20.34, and 23.70 min, respectively, are either very small peaks or not detected in the chromatogram (b), but they are clearly detected again in the chromatogram (c). It is also clear that several hydrophilic compounds with log K_{ow} of less than 3.0, for example, ethyl butyrate (RT: 6.44 min; log K_{ow}: 1.85), hexanol (RT: 11.41 min; log K_{ow}: 1.82), phenethyl alcohol (RT: 19.09 min; log K_{ow}: 1.57), and *cis*-β-methyl-γ-octalactone (RT: 19.76 min; log K_{ow}: 2.00), are more intense in the chromatogram (b) than in the chromatogram (a) and they are also clearly present in the chromatogram (c). Sequential SBSE thus gives the best enrichment over the entire polarity/volatility range.

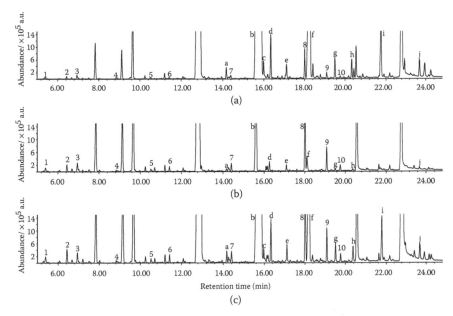

Figure 1.3 Total ion chromatograms of the single malt whiskey "B 12 years" obtained by (a) conventional SBSE without modifier, (b) conventional SBSE with salt addition (30% NaCl), and (c) sequential SBSE. 1. Ethyl isobutyrate (log K_{ow}: 1.77); 2. Ethyl butyrate (log K_{ow}: 1.85); 3. Ethyl isovalerate (log K_{ow}: 2.26); 4. 2-Heptanone (log K_{ow}: 1.73); 5. Furfuryl formate (log K_{ow}: 0.90); 6. Hexanol (log K_{ow}: 1.82); 7. Octanol (log K_{ow}: 2.81); 8. Phenethyl acetate (log K_{ow}: 2.57); 9. Phenethyl alcohol (log K_{ow}: 1.57); 10. γ-Nonalactone (log K_{ow}: 2.08). a. Ethyl nonanoate (log K_{ow}: 4.30); b. Ethyl decanoate (log K_{ow}: 4.79); c. Isoamyl octanoate (log K_{ow}: 5.24); d. Ethyl 9-decenoate (log K_{ow}: na); e. Decanol (log K_{ow}: 3.78); f. Ethyl dodecanoate (log K_{ow}: 5.78); g. Dodecanol (log K_{ow}: 4.77); h. Nerolidol (log K_{ow}: 5.68); i. Tetradecanol (log K_{ow}: 5.75); j. Farnesol (log K_{ow}: 5.77).

Figure 1.4 shows a comparison of relative intensities of selected compounds in the whiskey sample obtained by conventional SBSE without modifier, conventional SBSE with salt addition (30% NaCl), and sequential SBSE.

Analysis of beer

Odor compounds in beer cover a wide polarity range and are also present in a wide concentration range (from subnanogram per milliliter to high microgram per milliliter). SBSE has been successfully applied to the analysis of odor compounds in beer. Kishimoto et al. (2005) described the analysis of hop-derived terpenoids at subnanogram per milliliter to high microgram per milliliter levels, which provide distinctive aroma in beer, by SBSE-TD-GC–MS. Marsili, Laskonis, and Kenaan (2007) demonstrated

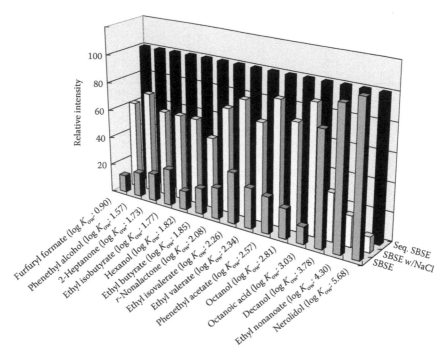

Figure 1.4 Comparison of relative intensities of selected compounds in the single malt whiskey "B 12 years" obtained by conventional SBSE without modifier, conventional SBSE with salt addition (30% NaCl), and sequential SBSE.

a method for analysis of off-flavor compounds, which were generated by heat- and light-abuse conditions, at subnanogram per milliliter to high microgram per milliliter levels based on SBSE-TD-GC–TOFMS and deconvolution software. Ochiai et al. (2003) described the analysis of stale-flavor aldehydes (e.g., E-2-nonenal and E, E-2,4-decadienal) at picogram per milliliter levels by SBSE with in situ derivatization using pentafluorobenzylhydroxylamine followed by TD-GC–MS. Horák et al. (2008) used SBSE with liquid desorption (LD) followed by GC–FID for the analysis of medium-chain fatty acids at submicrogram per milliliter levels, which play a major role for rancid/cheesy/sweaty off-flavor in beer as well as foam stability.

An example of the analysis of Japanese beer obtained by sequential SBSE-TD-GC–MS is shown in Figure 1.5. Although several hundred compounds were detected in the TIC, selected odor compounds, which have a wide polarity range, are identified in Figure 1.5, including isobutanoic acid (log K_{ow}: 1.00), valeric acid (log K_{ow}: 1.56), indole (log K_{ow}: 2.05), vinyl guaiacol (log K_{ow}: 2.24), α-terpineol (log K_{ow}: 3.33), decanoic acid (log K_{ow}: 4.02), dodecanoic acid (log K_{ow}: 5.00), and farnesol (log K_{ow}: 5.77). Nine odor

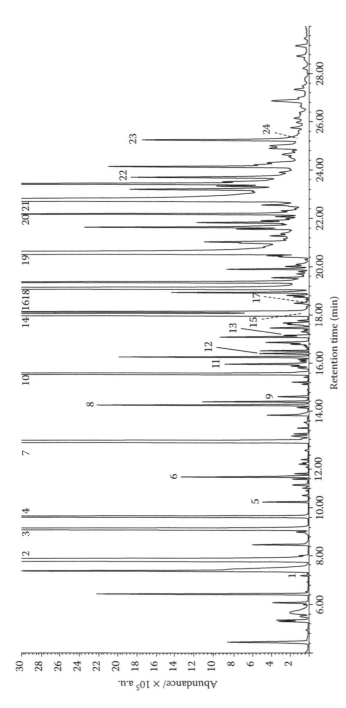

Figure 1.5 Total ion chromatograms of beer obtained by sequential SBSE-TD-GC–MS. 1. Hexanal (log K_{ow}: 1.80); 2. Isoamyl acetate (log K_{ow}: 2.26); 3. Isoamyl alcohol (log K_{ow}: 1.26); 4. Ethyl hexanoate (log K_{ow}: 2.83); 5. Hexyl acetate (log K_{ow}: 2.83); 6. Hexanol (log K_{ow}: 1.82); 7. Ethyl octanoate (log K_{ow}: 3.81); 8. Linalool (log K_{ow}: 3.38); 9. Iso-butanoic acid (log K_{ow}: 1.00); 10. Ethyl decanoate (log K_{ow}: 4.79); 11. Valeric acid (log K_{ow}: 1.56); 12. α-Terpineol (log K_{ow}: 3.33); 13. Citronellol (log K_{ow}: 3.56); 14. Phenethyl acetate (log K_{ow}: 2.57); 15. β-Damascenone (log K_{ow}: 4.21); 16. Hexanoic acid (log K_{ow}: 2.05); 17. Guaiacol (log K_{ow}: 1.57); 18. Phenethyl alcohol (log K_{ow}: 1.57); 19. Octanoic acid (log K_{ow}: 3.03); 20. Vinyl guaiacol (log K_{ow}: 2.24); 21. Decanoic acid (log K_{ow}: 4.02); 22. Farnesol (log K_{ow}: 5.77); 23. Dodecanoic acid (log K_{ow}: 5.00); 24. Indole (log K_{ow}: 2.05).

Table 1.2 Selected Odor Compound Concentrations in Beer
Obtained by Sequential SBSE-TD-GC–MS

Compound	log K_{ow}[a]	Linearity (r^2)	Concentration (ng/mL)	RSD (%), $n = 5$
Guaiacol	1.34	0.9981[b]	4.4	9.7
Hexanal	1.80	0.9917[b]	4.4	5.4
Hexanol	1.82	0.9958[b]	20	5.1
Phenethyl acetate	2.57	0.9940[c]	730	2.0
Ethyl hexanoate	2.83	0.9926[c]	270	1.5
Linalool	3.38	0.9961[b]	27	3.9
Citronellol	3.56	0.9981[b]	3.9	6.5
Ethyl octanoate	3.81	0.9940[c]	530	4.6
β-Damascenone	4.21	0.9919[b]	6.4	8.2

[a] The log K_{ow} values are calculated with SRC-KOWWIN software.
[b] Linearity of standard addition calibration curve between 5 and 40 ng/mL.
[c] Linearity of standard addition calibration curve between 200 and 2000 ng/mL.

compounds were determined using a standard addition calibration method (Table 1.2). Very good linearity ($r^2 > 0.9917$) and repeatability (RSD% < 10%, $n = 5$) were obtained for all analytes. Six minor constituents, for example, guaiacol (log K_{ow}: 1.34), hexanal (log K_{ow}: 1.80), hexanol (log K_{ow}: 1.82), linalool (log K_{ow}: 3.38), citronellol (log K_{ow}: 3.56), and β-damascenone (log K_{ow}: 4.21), were in the range of 4.4–27 ng/mL. Three major constituents, phenethyl acetate (log K_{ow}: 2.57), ethyl hexanoate (log K_{ow}: 2.83), and ethyl octanoate (log K_{ow}: 3.81), were in the range of 270–730 ng/mL. Figure 1.6 shows the mass chromatogram (m/z 124) of a nonspiked beer and of a beer spiked at 20 ng/mL guaiacol. Figure 1.7 shows the measured mass spectrum of guaiacol in the nonspiked beer (at 4.4 ng/mL) and a National Institute of Standards and Technology library mass spectrum of guaiacol.

Sequential SBSE with PDMS stir bar and EG Silicon stir bar

Stir bar sorptive extraction with new phase stir bar

In the past decade, PDMS has been the only available extraction phase on commercial stir bars. Several types of extraction phase with different polarities have been proposed to extend the utility of the PDMS phase in the immersion mode. Lambert et al. (2005) developed a biocompatible stir bar coated with alkyl-diol-silica restricted access material as extraction phase for analysis of caffeine and metabolites in biological fluids. Huang et al. developed a new phase based on monolithic materials, for example, polymethacrylic acid stearyl ester-ethylene dimethacrylate derivative and polyvinylpyridine-ethylene

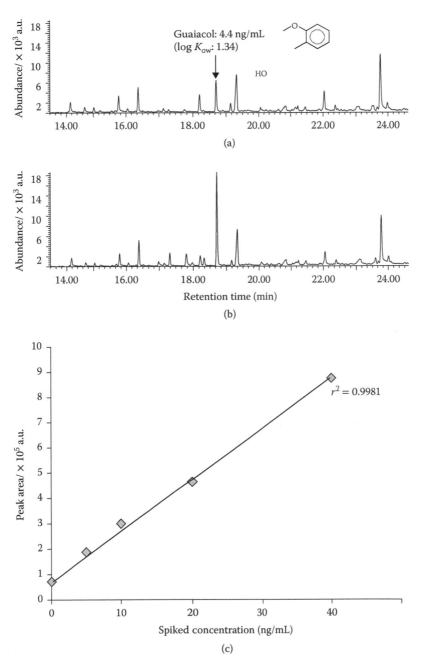

Figure 1.6 Mass chromatograms (*m/z* 124) for guaiacol (log K_{ow}: 1.34) in (a) non-spiked and (b) spiked (20 ng/mL) beer sample. (c) Standard addition calibration curve of guaiacol in the concentration range between 5 and 40 ng/mL.

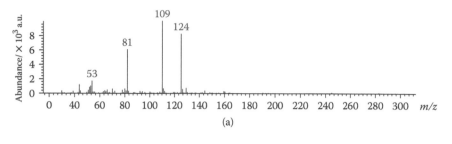

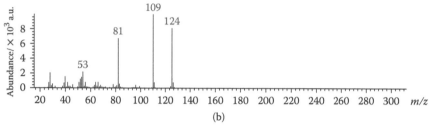

Figure 1.7 Measured mass spectra of guaiacol (4.4 ng/mL) in a beer sample obtained by (a) sequential SBSE-TD-GC–MS, and (b) a National Institute of Standards and Technology library mass spectra of guaiacol.

dimethacrylate derivative, to analyze steroid sex hormones in urine (Huang, Yuan, and Huang 2008) and phenols in water (Huang, Qiu, and Yuan 2008). These new phase stir bars are used with LD-HPLC analysis because of lack of thermal stability to allow TD and GC analysis.

Recently, a new phase stir bar, which has PEG-modified silicon coating (EG Silicon), was developed for TD-GC analysis (Gerstel). The EG Silicon stir bar can show good performance in terms of bleeding, repeatability, and robustness as is the case of the PDMS stir bar. Compared with the PDMS stir bar, the EG Silicon stir bar can provide higher recovery for polar (hydrophilic) compounds, specifically with salt addition.

Extraction efficiency of odor compounds

In sequential SBSE, the first extraction is performed with unmodified sample for the compounds with high K_{ow} (log K_{ow} > 4.0), the second extraction is performed with a modified sample solution (containing 30% NaCl) for the compounds with low and medium K_{ow} (log K_{ow} < 3.0–4.0). To obtain more uniform enrichment over the entire polarity range for odor compounds in aqueous samples, sequential SBSE with the first extraction using the PDMS stir bar for unmodified sample and the second extraction using the EG Silicon stir bar for modified sample (containing 30% NaCl) were performed (PDMS–EG Silicon combination). Figure 1.8 shows a comparison of the TIC of the single malt whiskey "L 10

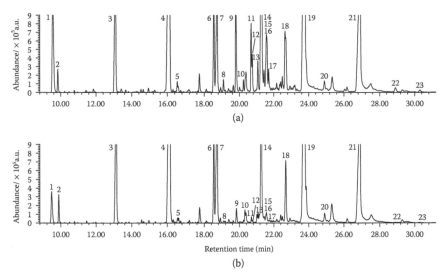

Figure 1.8 Total ion chromatograms (TIC) of the single malt whiskey "L 10 years" obtained by sequential SBSE with (a) the PDMS–EG Silicon combination and (b) the PDMS–PDMS combination. 1. Isoamyl alcohol (log K_{ow}: 1.26); 2. Ethyl hexanoate (log K_{ow}: 2.83); 3. Ethyl octanoate (log K_{ow}: 3.81); 4. Ethyl decanoate (log K_{ow}: 4.79); 5. Diethyl succinate (log K_{ow}: 1.39); 6. Phenethyl acetate (log K_{ow}: 2.57); 7. Hexanoic acid (log K_{ow}: 2.05); 8. Guaiacol (log K_{ow}: 1.34); 9. Phenethyl alcohol (log K_{ow}: 1.57); 10. 2-Methoxy-4-methyl phenol (log K_{ow}: 1.88); 11. *o*-Cresol (log K_{ow}: 2.06); 12. Phenol (log K_{ow}: 1.51); 13. 4-Ethyl guaiacol (log K_{ow}: 2.38); 14. Octanoic acid (log K_{ow}: 3.03); 15. *p*-Cresol (log K_{ow}: 2.06); 16. 2,4-Dimethyl phenol (log K_{ow}: 2.61); 17. *m*-Cresol (log K_{ow}: 2.06); 18. 2-Ethyl phenol (log K_{ow}: 2.55); 19. Decanoic acid (log K_{ow}: 4.02); 20. Farnesol (log K_{ow}: 5.77); 21. Dodecanoic acid (log K_{ow}: 5.00); 22. Vanillin (log K_{ow}: 1.05); 23. Ethyl vanillate (log K_{ow}: 2.31).

years" (fivefold diluted with natural water) between (a) the PDMS–EG Silicon combination and (b) the PDMS–PDMS combination. Figure 1.9 shows a comparison of the mass chromatogram (*m/z* 151) between (a) the PDMS–EG Silicon combination and (b) the PDMS–PDMS combination. It is clearly observed that several hydrophilic compounds with log K_{ow} of less than 2.5, specifically for phenolic compounds, for example, vanillin (log K_{ow}: 1.05), guaiacol (log K_{ow}: 1.34), phenol (log K_{ow}: 1.51), cresol (log K_{ow}: 2.06), ethyl vanillate (log K_{ow}: 2.38), and 4-ethyl guaiacol (log K_{ow}: 2.38), are more intense in the chromatogram (a) than in the chromatogram (b). These phenolic compounds are responsible for specific flavor from the peated malt whiskey. Table 1.3 shows a comparison of extracted amounts for the selected compounds between the PDMS–EG Silicon combination and the PDMS–PDMS combination. Although the enrichment factors (PDMS–EG Silicon/PDMS–PDMS) for the compounds with log K_{ow} of more than 2.5 were in the range of 0.96–1.2 (the extraction

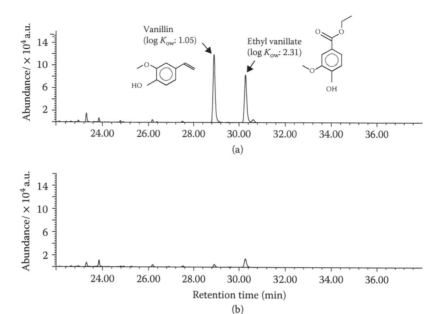

Figure 1.9 Mass chromatograms (m/z 151) for vanillin (log K_{ow}: 1.05) and ethyl vanillate (log K_{ow}: 2.31) in the single malt whiskey "L 10 years" obtained by sequential SBSE with (a) the PDMS–EG Silicon combination and (b) the PDMS–PDMS combination.

Table 1.3 Comparison of Extracted Amounts for the Selected Compounds in Whiskey

		Extracted amount (ng)		
Compound	log K_{ow}[a]	PDMS–PDMS[b]	PDMS–EG Silicon[c]	Enrichment factor[d]
Vanillin	1.05	14	400	29
Guaiacol	1.34	45	340	7.6
Phenethyl alcohol	1.57	380	3700	10
Hexanol	1.82	25	110	4.4
2-Methyl phenol	2.06	110	1500	14
4-Ethyl guaiacol	2.38	210	550	2.6
Phenethyl acetate	2.57	1800	2200	1.2
Ethyl hexanoate	2.83	640	640	1.0
Ethyl octanoate	3.81	4600	4400	0.96

[a] The log K_{ow} values are calculated with SRC-KOWWIN software.
[b] The first SBSE using the PDMS stir bar for unmodified sample, and the second SBSE using the PDMS stir bar for modified sample (containing 30% NaCl).
[c] The first SBSE using the PDMS stir bar for unmodified sample, and the second SBSE using the EG Silicon stir bar for modified sample (containing 30% NaCl).
[d] PDMS–EG Silicon/PDMS–PDMS.

efficiencies for these compounds are almost the same), the enrichment factors for several hydrophilic compounds in the whiskey sample are much higher, for example, for vanillin (log K_{ow}: 1.05), guaiacol (log K_{ow}: 1.34), and phenethyl alcohol (log K_{ow}: 1.57), the enrichment factors are 29, 7.6, and 10, respectively.

Figure 1.10 shows a comparison of the NPD chromatogram of a canned coffee (ready-to-drink) between (a) the PDMS–EG Silicon combination and (b) the PDMS–PDMS combination. It is clear that certain types of nitrogen compounds, for example, 2-acetyl pyrrole (log K_{ow}: 0.56), 2-formyl pyrrole (log K_{ow}: 0.60), 1H-pyrrole (log K_{ow}: 0.88), 1-methyl-2-acetyl pyrrole (log K_{ow}: 1.11), 1-methyl-2-formyl pyrrole (log K_{ow}: 1.14), methyl pyrrole (log K_{ow}: 1.43), indole (log K_{ow}: 2.05), and furfuryl pyrrole (log K_{ow}: 2.50), are significantly better recovered from the canned coffee sample whereas other nitrogen compounds are similarly recovered. Figure 1.11 shows a comparison of peak areas of selected nitrogen compounds obtained by

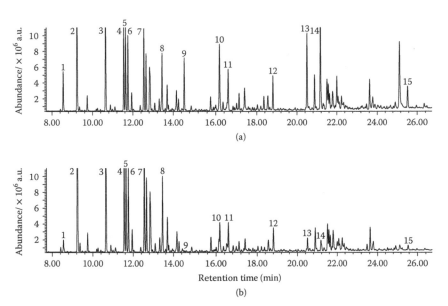

Figure 1.10 GC–NPD chromatograms of a canned coffee (ready-to-drink) obtained by sequential SBSE with (a) the PDMS–EG Silicon and (b) the PDMS–PDMS combination. 1. Methyl pyrrole (log K_{ow}: 1.43); 2. Pyridine (log K_{ow}: 0.80); 3. Methyl pyrazine (log K_{ow}: 0.49); 4. 2,5-Dimethyl-pyrazine (log K_{ow}: 1.03); 5. 2,6-Dimethyl-pyrazine (log K_{ow}: 1.03); 6. Ethyl pyrazine (log K_{ow}: 0.98); 7. 2-Ethyl-6-methyl-pyrazine (log K_{ow}: 1.53); 8. 2-Ethyl-3,5-dimethyl-pyrazine (log K_{ow}: 2.07); 9. 1H-Pyrrole (log K_{ow}: 0.88); 10. 1-Methyl-2-formyl pyrrole (log K_{ow}: 1.14); 11. 1-Methyl-2-acetyl pyrrole (log K_{ow}: 1.11); 12. 4-Furfuryl pyrrole (log K_{ow}: 2.50); 13. 2-Acetyl pyrrole (log K_{ow}: 0.56); 14. 2-Formyl pyrrole (log K_{ow}: 0.60); 15. Indole (log K_{ow}: 2.05).

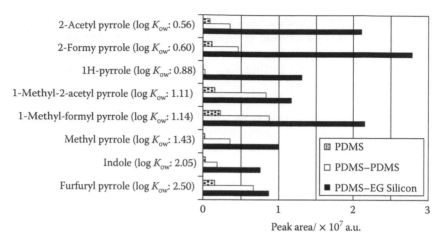

Figure 1.11 Comparison of selected nitrogen compounds in a canned coffee (ready-to-drink) obtained by conventional SBSE, sequential SBSE with the PDMS–PDMS combination, and sequential SBSE with the PDMS–EG Silicon combination.

single SBSE with the PDMS stir bar, the PDMS–PDMS combination, and the PDMS–EG Silicon combination.

The results illustrated in Figures 1.8 through 1.11 impressively reflect the affinity of the EG Silicon stir bar for hydrophilic compounds, for example, phenol, alcohol, and pyrrole, which can form hydrogen bonding. Also, the results show that sequential SBSE with the PDMS–EG Silicon combination can provide more uniform enrichment of a wide range of odor compounds in aqueous samples.

Conclusion

The present studies demonstrated the practicability and the feasibility of sequential SBSE for the analysis of odor compounds in aqueous samples. With sequential SBSE, a wide range of odor compounds with different polarities can be uniformly extracted and enriched, while the negative effect of salt addition on recovery of compounds with high K_{ow} (log K_{ow} > 4.0) is eliminated. Also, quantitative analysis of a wide range of odor compounds in an aqueous sample (e.g., beer) could be performed with sequential SBSE using a standard addition calibration method. Moreover, by using sequential SBSE with the PDMS–EG Silicon combination, much better extraction efficiency for hydrophilic compounds (e.g., phenol, alcohol, and pyrrole) could be obtained compared with sequential SBSE with the PDMS–PDMS combination as well as conventional SBSE.

Acknowledgments

I thank Kikuo Sasamoto of Gerstel K. K., Edward Pfannkoch of Gerstel Inc., Yunyun Nie, and Eike Kleine-Benne of Gerstel GmbH & Co. KG. for their kind support in preparing this manuscript. Also I wish to express appreciation to Dr. Frank David of Research Institute for Chromatography for his valuable comments.

References

Arthur, C. L., and J. Pawliszyn. 1990. Solid phase microextraction with thermal desorption using fused silica optical fibers. *Anal Chem* 62:2145–8.

Baltussen, E., P. Sandra, F. David, and C. A. Cramers. 1999. Stir bar sorptive extraction (SBSE), a novel extraction technique for aqueous samples: Theory and principles. *J Microcolumn Sep* 11(10):737–47.

David, F., and P. Sandra. 2007. Stir bar sorptive extraction for trace analysis. *J Chromatogr A* 1152:54–69.

Horák, T., J. Čulík, M. Jurková, P. Čejka, and V. Kellner. 2008. Determination of free medium-chain fatty acids in beer by stir bar sorptive extraction. *J Chromatogr A* 1196–7:96–9.

Huang, X. J., N. N. Qiu, and D. X. Yuan. 2008. Direct enrichment of phenols in lake and sea water by stir bar sorptive extraction based on poly (vinylpyridine-ethylene dimethacrylate) monolithic material and liquid chromatographic analysis. *J Chromatogr A* 1194:134–8.

Huang, X., D. Yuan, and J. B. L. Huang. 2008. Determination of steroid sex hormones in urine matrix by stir bar sorptive extraction based on monolithic material and liquid chromatography with diode array detection. *Talanta* 75:172–7.

Kishimoto, T., A. Wanikawa, N. Kagami, and K. Kawatsura. 2005. Analysis of hop-derived terpenoids in beer and evaluation of their behavior using the stir bar sorptive extraction method with GC–MS. *J Agric Food Chem* 53:4701–7.

Lambert, J.-P., W. M. Mullett, E. Kwong, and D. Lubda. 2005. Stir bar sorptive extraction based on restricted access material for the direct extraction of caffeine and metabolites in biological fluids. *J Chromatogr A* 1075:43–9.

Leon, V. M., B. Alvarez, M. A. Cobollo, S. Munoz, and I. Valor. 2003. Analysis of 35 priority semivolatile compounds in water by stir bar sorptive extraction-thermal desorption-gas chromatography-mass spectrometry. I. Method optimization. *J Chromatogr A* 999:91–101.

Marsili, R. T., L. C. Laskonis, and C. Kenaan. 2007. Evaluation of PDMS-based extraction techniques and GC–TOF-MS for the analysis of off-flavor chemicals in beer. *J Am Soc Brew Chem* 65(3):129–37.

Marsili, R., J. Szleszinski, D. Brieter, and M. Drake. 2009. Sequential SBSE GC–MS analysis of trihaloanisoles & other off-flavor chemicals in casein powder. Paper presented at the NCFST- Method Innovations in Melamine and Other Food Contaminant Analysis Workshop, Summit-Argo, Illinois.

Meylan, W. M., and P. H. Howard. 1995. Atom/fragment contribution method for estimating octanol-water partition coefficients. *J Pharm Sci* 84:83–92.

Nakamura, S., and S. Daishima. 2005. Simultaneous determination of 64 pesticides in river water by stir bar sorptive extraction and thermal desorption-gas chromatography-mass spectrometry. *Anal Bioanal Chem* 382:99–107.

Ochiai, N., K. Sasamoto, S. Daishima, A. C. Heiden, and A. Hoffmann. 2003. Determination of stale-flavor carbonyl compounds in beer by stir bar sorptive extraction with in-situ derivatization and thermal desorption–gas chromatography–mass spectrometry. *J Chromatogr A* 986:101–10.

Ochiai, N., K. Sasamoto, H. Kanda, and E. Pfannkoch. 2008. Sequential stir bar sorptive extraction for uniform enrichment of trace amounts of organic pollutants in water samples. *J Chromatogr A* 1200:72–9.

chapter two

Selectable one-dimensional or two-dimensional gas chromatography–mass spectrometry

Nobuo Ochiai

Contents

Introduction

Gas chromatography–olfactometry (GC–O) is a valuable method for the selection of odor compounds from a complex mixture (Blank 2002). In particular, GC–O in combination with mass spectrometry (MS; GC–O/MS) allows not only the evaluation of odor compounds but also their identification with mass spectral information. However, many key odor compounds can occur at very low concentrations. Therefore, identification of odor compounds remains a hard task even with GC–O/MS because some compounds coelute with other analytes or the sample matrix, which leads to difficulties when correlating the detected aroma with the correct compound. Multidimensional (MD) GC, more specifically, two-dimensional (^{2}D) GC with simultaneous olfactometry and mass spectrometric detection, extends the identification capability and separation resolution of GC. There are two established ^{2}D GC approaches: (1) conventional heart-cutting ^{2}D GC (GC–GC) (Bertsch 1990) and (2) comprehensive ^{2}D GC (GCxGC) (Adahchour et al. 2006). The former approach is commonly used in target analysis of specific compounds in a sample. Heart-cutting ^{2}D GC–MS with olfactometry can significantly improve the identification capability as well as the resolution of complex regions (Nitz, Kollmannsberger, and Drawert 1989; Pfannkoch and Whitecavage 2005). In contrast, the comprehensive ^{2}D GC approach is mainly used in exhaustive analysis of a sample for total profiling. In the comprehensive approach, after a single introduction to the first column the entire sample is subjected to the two different separations (Adahchour et al. 2006). However, since GC run time in the second dimension of GCxGC is an order of magnitude faster than that in the first dimension (typically 3–5 s), it is very difficult to correlate a given odor perceived in GCxGC–O analysis to the peaks in the second dimension separation at this stage even after the first publication of GCxGC–O for the analysis of perfume (Zellner et al. 2007).

This chapter discusses a novel heart-cutting ^{2}D GC–MS system referred to as a selectable ^{1}D or ^{2}D GC–MS (^{1}D/^{2}D GC–MS) with simultaneous olfactometry or element-specific detection for the analysis of odor compounds in complex samples. The combination of ^{1}D/^{2}D GC–O/MS with preparative fraction collection (PFC) for the enrichment of odor compounds is also discussed. In addition, the practical applications of ^{1}D/^{2}D GC–MS with simultaneous detection are demonstrated with the use of novel sample preparation–introduction techniques, for example, stir bar sorptive extraction (SBSE)–thermal desorption (TD) (David and Sandra 2007) and "large volume full evaporation technique" (LVFET)–TD (Hoffmann, Lerch, and Hudewenz 2009), which enable an automated, miniaturized, and solvent-free method.

Heart-cutting two-dimensional gas chromatography

A modern heart-cutting ²D GC system uses electronic pneumatic control (EPC) technology, which uses electronic mass-flow controllers and electronic proportional back-pressure or forward-pressure regulator, and a microfluidic device (MFD) based on the Deans principle (Deans 1968). The combination of EPC technology and MFD has eliminated many deficiencies encountered with the classical MDGC system using the Deans principle (e.g., reproducibility of cutting and retention time [RT] stability). Several companies offer MDGC systems using EPC technology and MFD. Recently, several researchers (Luong et al. 2008; Feyerhem et al. 2007; David and Klee 2009; Sasamoto and Ochiai 2008) reported a new heart-cutting ²D GC–MS approach with low thermal mass (LTM)–GC (Luong et al. 2007). The LTM–GC is a modular GC system that consists of a standard fused-silica capillary column with separate heating, temperature-sensing, and insulating elements. The LTM–GC system can provide not only rapid heating and cooling but also independent temperature control for MDGC. The use of LTM–GC, in addition to EPC technology and MFD, has revitalized the conventional heart-cutting ²D GC–MS technique and can provide a high degree of flexibility in ²D GC–MS analysis, for example, a combination of heart-cut and backflushing, independent temperature control of each chromatographic dimension to maximize separation power, thermal focusing at the head of the second dimension column, fast temperature-programming rates, and rapid cooldown.

Selectable one-dimensional or two-dimensional gas chromatography–mass spectrometry with simultaneous detection

Figure 2.1 shows the typical schematic of a conventional heart-cutting ²D GC–MS system. The ¹D mode is to have the sample pass through the ¹D GC column to the monitor detector such as a flame ionization detector (FID). The switching device (e.g., an MFD Deans switch) is applied to the interface to transfer the certain peaks eluted from the ¹D GC column to the ²D GC column for MS detection. For heart-cutting ²D GC–MS analysis one usually requires a dedicated ²D GC–MS system such as the one described in Figure 2.1, which is then not available for routine ¹D GC–MS analysis as it is already configured. Therefore, one must have two different GC–MS systems (¹D GC–MS and ²D GC–MS) or one has to set up once again a single GC–MS system for heart-cutting ²D GC–MS applications. Also, the ¹D chromatogram obtained by heart-cutting ²D GC–MS is normally generated with a monitor detector such as an FID. Thus, selection and confirmation

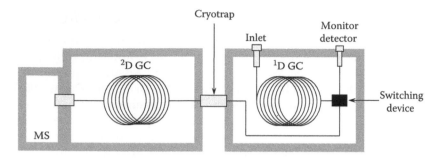

Figure 2.1 Schematic flow diagram of a conventional heart-cutting two-dimensional gas chromatography–mass spectrometry (^{2}D GC–MS) system.

of the heart-cut region for heart-cutting ^{2}D GC–MS analysis must be done without mass spectral information. In this case, preliminary tests with the injection of an authentic compound or known compound prior to each heart-cutting ^{2}D GC–MS application are required.

In order to overcome these bottlenecks, Sasamoto and Ochiai (2010) developed a novel selectable ^{1}D or ^{2}D GC–MS (^{1}D/^{2}D GC–MS) system using MFD and dual LTM–GC for the simple and fast operation of both ^{1}D GC–MS and ^{2}D GC–MS with simultaneous detection using a single GC–MS system. The selection of ^{1}D/^{2}D GC–MS operation is easily performed by a mouse click, without any change in the instrumental setup. Also, this system can eliminate preliminary analysis with a "monitor FID" for the selection of the heart-cut region and can provide a "monitor total ion chromatogram (TIC)" for the ^{1}D column separation on ^{2}D GC–MS analysis. Moreover, simultaneous mass spectrometric and olfactometry or element-specific detection can be performed for both the ^{1}D column separation and the ^{2}D column separation under a constant split ratio.

Instrumentation

Stir bars coated with 24 μL of polydimethylsiloxane (PDMS; Twister) were obtained from Gerstel, Inc. (Gerstel, Mülheim an der Ruhr, Germany). For SBSE, 10-mL headspace vials with screw cap containing polytetrafluoro-ethylene (PTFE)-coated silicon septa (Gerstel) were used. The SBSE process was performed with a multiposition magnetic stirrer (20 positions) from Global Change (Tokyo, Japan). Automated LVFET was performed using a dynamic headspace module (DHS; Gerstel), which was subsequently followed by TD–GC–MS analysis using a TD unit (TDU) equipped with a MultiPurpose Sampler (MPS) 2 autosampler and a Cooled Inlet System (CIS) 4 programmed temperature vaporization (PTV) inlet (Gerstel), dual LTM–GC system (Agilent Technologies, California), a nitrogen phosphorus detector (NPD; Agilent), a 5380 Pulsed Flame Photometric Detector (PFPD,

OI Analytical, Texas) installed on an Agilent 7890 gas chromatograph (host GC) with a 5975C mass-selective detector. The LTM–GC system consists of dual wide format column modules (5 inches; 1 inch = 2.54 cm), LTM-heated transfer lines, cooling fan, temperature controller, power supply, and a specially constructed GC door. The GC was equipped with a switching device that consists of an MFD Deans switch (Agilent), an MFD three-way splitter with makeup gas line (Agilent) and restrictor. The switching device is controlled using a pressure control module (PCM). For simultaneous MS and dual selective detections, an additional two-way splitter is used. A cryotrap system (CTS2; Gerstel) was used for the linear retention indices (LRIs) study.

Stir bar sorptive extraction

The SBSE method is a miniaturized extraction method, which is a simple, solventless technique allowing extraction and concentration in a single step. This method has been successfully applied to the determination of organic compounds in various sample matrices, for example, water, soil, food, and biological fluid. Also, SBSE provides enhanced sensitivity because the extracted fraction (on a stir bar) can be introduced quantitatively into a GC system by TD. Several authors indicate that the SBSE method allows high recovery and extremely low limit of detection (LOD) at the sub-nanogram per liter level, particularly for solutes having hydrophobic characteristics (David and Sandra 2007).

Prior to use, the stir bars were conditioned for 30 min at 250°C in a flow of helium. For SBSE, 1 mL of whiskey sample and 4 mL of natural water were transferred to 10-mL headspace vials. A stir bar was added and the vial was sealed with a screw cap. The SBSE of several samples was performed simultaneously at room temperature (24°C) for 60 min while stirring at 1500 rpm. After extraction, the stir bar was removed using a forceps, dipped briefly in Milli-Q water, dried with a lint-free tissue, and placed in a glass TD liner. The glass liner was placed in the TDU. No further sample preparation was necessary. The analytical conditions are summarized in Table 2.1.

Reconditioning of stir bars was done after use by soaking in Milli-Q purified water and acetonitrile for 24 h each; stir bars were then removed from the solvent and dried on a clean surface at room temperature for 1 h. Finally, the stir bars were thermally conditioned for 30 min at 250°C in a flow of helium. Typically, more than 30 extractions can be performed with the same stir bar.

Large volume full evaporation technique

The technique of introducing a small volume of sample and vaporizing the analytes in the headspace vial completely, without having to rely on establishing equilibrium between two phases, is called a full evaporation

Table 2.1 Analytical Conditions

SBSE	
Sample volume	5 mL (fivefold diluted whiskey) in 10-mL vial
Stir bar	Twister (10-mm length × 500-µm thickness)
Extraction	1500 rpm, 1 h
LVFET	
Sample volume	100 µL in 10-mL vial
Sample temperature	80°C
Trap	Tenax TA packed TDU tube
Trap temperature	40°C
Purge flow rate	100 mL/min
Purge volume	3 L
TDU	
TD temperature	30°C (0.5 min)–12°C·s^{-1}–250°C or 280°C (3 min)
Desorption mode	Splitless
PTV	
PTV liner	Tenax TA packed CIS liner
Injection	0.2-min solvent vent (50 mL/min) split 1:1
PTV temperature	10°C (0.5 min)–12°C/s–250°C (hold)
Selectable ^{1}D/^{2}D GC	
First column	DB-Wax (10 m × 0.18 mm i.d. [internal diameter] × 0.30 µm d_f [film thickness])
	DB-Wax (30 m × 0.25 mm i.d. × 0.25 µm d_f)
First column flow rate	3.5 mL/min
First column temperature	40°C (2 min)–10°C/min–240°C (hold)
Second column	DB-1 (10 m × 0.18 mm i.d.
	DB-5 (10 m × 0.18 mm i.d. × 0.40 µm d_f)
Second column flow rate	4.5 mL/min
Second column temperature	40°C–10°C/min–280°C (hold)
Host GC Oven	
	250°C
MS	
Ionization	Electron ionization
Scan range	*m/z*: 29–300
Scan speed	2.83 scans·s^{-1}

Table 2.1 Analytical Conditions (*Continued*)

Selectable ^{1}D/^{2}D GC–O/MS	
Split ratio (olfactometry:MS)	2:1
Transfer-line temperature	250°C
Selectable ^{1}D/^{2}D GC–PFPD/NPD/MS	
Split ratio (PFPD:NPD:MS)	1:1:1
PFPD temperature	250°C
PFPD mode	S-trace
NPD temperature	325°C

technique (FET) (Markelov and Guzouwski 1993; Markelov and Bershevits 2001). The FET is usually performed with not full evaporation of the condensed phase but with near-complete transfer of the analytes into the vapor phase. However, the application of FET for more hydrophilic compounds in an aqueous sample is limited because the maximum mass of water that can be evaporated in a typical size of headspace vial (e.g., 22.5 mL) at an elevated temperature (e.g., 100°C) is less than 14 mg (Markelov and Guzouwski 1993). Therefore, more hydrophilic compounds in a larger size of aqueous sample (e.g., 100 µL) cannot transfer to the vapor phase in conventional headspace vials (sizes of 10–22.5 mL). In order to apply FET to the analysis of more hydrophilic compounds as well as hydrophobic compounds, Hoffmann, Lerch, and Hudewenz (2009) revitalized the concept of FET using DHS. This technique is called LVFET.

For LVFET, typically 100 µL of the aqueous sample (e.g., whiskey) is placed in a 10-mL headspace vial, which is thermostated to 80°C allowing the aqueous matrix to vaporize completely with nitrogen gas purge at a flow rate of 100 mL/min while most of the low-volatile matrix is left behind. Thanks to the complete evaporation of the aqueous sample with the continuous purging of gas at elevated temperature, more hydrophilic compounds can now be transferred to the vapor phase. Analytes in the purged headspace are trapped onto an adsorbent bed in a TDU glass tube. The analytical conditions are summarized in Table 2.1.

Selectable one-dimensional or two-dimensional gas chromatography–mass spectrometry with simultaneous olfactometry or element-specific detection

Figure 2.2 shows a flow diagram of the ^{1}D/^{2}D GC–MS system with simultaneous olfactometry and/or element-specific detection. The system consists of an inlet, dual LTM–GC, a combination of an MFD Deans switch (Deans) and an MFD three-way splitter with makeup gas line (splitter),

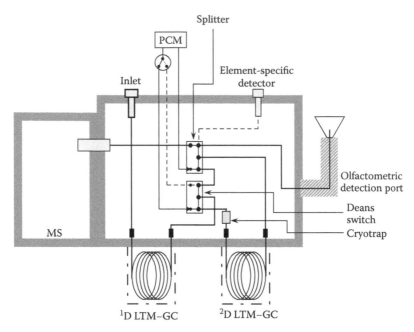

Figure 2.2 Schematic flow diagram of a selectable ¹D/²D GC–MS system with simultaneous detection system. PCM = pressure control module.

a host GC oven, an olfactometry detection port (ODP) and/or selective detectors, and an mass spectrometer. This system has three thermal zones: (1) LTM–GC1, (2) LTM–GC2, and (3) the host GC oven. Figure 2.3 shows a simplified flow diagram of the system. For ¹D GC–MS analysis (Figure 2.3a), compounds that are injected from an inlet are separated on LTM–GC1 and then transferred to the splitter through the restrictor. At the same time, carrier gas provided by the PCM keeps the LTM–GC2 clean and free from chromatographic interferences. Outlet flow from the restrictor and LTM–GC1 are merged at the splitter, which splits it to the MS and the D1. The split ratio is kept constant during the GC run by controlling a pressure at the splitter under a constant host GC oven temperature (250°C) with temperature programming on LTM–GC1 and LTM–GC2. Compounds of interest, which are located by the ODP and/or the selective detectors, are identified with spectral information at the same RT on the TIC. In case of coelution interference on mass spectral identification, the spectral information can work as a "monitor TIC" for the selection of a heart-cut region. Two-dimensional GC–MS analysis can be performed just after ¹D GC–MS analysis without any change in the instrumental setup. For ²D GC–MS analysis, ¹D GC–MS separation is performed until the heart-cutting is done, and then the heart-cut region is transferred onto LTM–GC2 (Figure 2.3b). In most cases, the transferred compounds are thermally focused at the head of LTM–GC2, which is kept at an initial temperature (typically, 40°C); therefore, no cryotrap is needed for a single heart-cutting. In the case of multiple heart-cutting or an LRI study, a cryotrap must be placed before LTM–GC2. After the transfer of

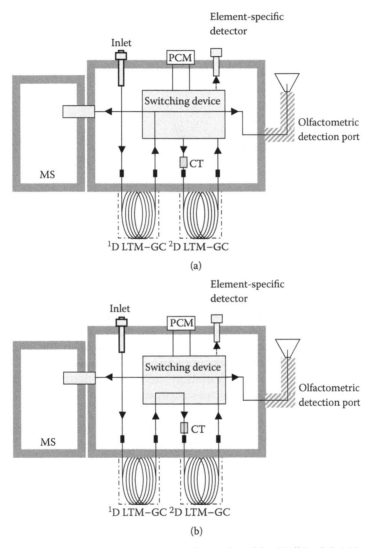

Figure 2.3 Schematic flow diagram of a selectable ¹D/²D GC–MS system with simultaneous detection system. (a) ¹D GC–MS analysis. (b) Heart-cutting. (c) ²D GC–MS analysis and ¹D GC backflush. CT = cryotrap and PCM = pressure control module.

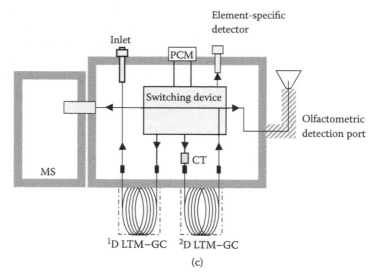

Figure 2.3 *(Continued)*

the heart-cut region (Figure 2.3c) the remaining sample on LTM–GC1 can be effectively backflushed by setting the inlet pressure to 10 kPa, and then the temperature program for LTM–GC2 starts. Finally, simultaneous MS and selective detection can be performed for the compounds of interest that are separated on LTM–GC2.

The calculation procedure for analytical conditions of the ^{1}D/^{2}D GC–MS with simultaneous detection, for example, all parameters, flow rate, and restrictor dimensions, is described elsewhere in the literature (Sasamoto and Ochiai 2010); this procedure is a complicated and time-consuming process. For this calculation, dedicated calculation software called ^{1}D/^{2}D Sync software is commercially available (Gerstel).

Evaluation of simultaneous detection in selectable one-dimensional or two-dimensional gas chromatography–mass spectrometry

With the selectable ^{1}D/^{2}D GC–MS system, simultaneous detection using the MS (vacuum) and other detectors (atmospheric pressure) is possible for both the ^{1}D GC and the ^{2}D GC approaches under constant split ratio and flow rates. Figure 2.4 shows both the ^{1}D and the ^{2}D TICs and the corresponding FID chromatograms of lemon oil obtained by ^{1}D/^{2}D GC–FID/MS. A heart-cut in the RT 11.70–12.07 min (22 s) region was performed to transfer coeluted peaks (at least three compounds) to the second dimension. The transferred peaks were thermally focused at the head of the

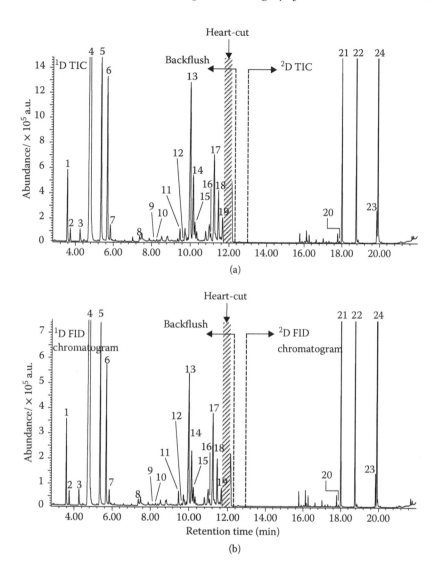

Figure 2.4 One-/two-dimensional (¹D/²D) total ion chromatogram (TIC) and ¹D/²D flame ionization detector (FID) chromatogram obtained by selectable ¹D/²D GC–FID/MS for a lemon oil. (a) ¹D/²D TIC. (b) ¹D/²D FID chromatogram. In the figure, ¹D column: DB-Wax, 10 m × 0.18 mm i.d. × 0.30 μm d_f, 40°C–10°C/min–240°C; ²D column: DB-1, 10 m × 0.18 mm i.d. × 0.40 μm d_f, 40°C–10°C/min–240°C; 1. β-pinene; 2. sabinene; 3. myrcene; 4. limonene; 5. γ-terpinene; 6. *p*-cymene; 7. terpinolene; 8. nonanal; 9. *cis*-limonene oxide; 10. *trans*-limonene oxide; 11. linalool; 12. nonylacetate; 13. *trans*-α-bergamotene; 14. β-caryophyllene; 15. 4-terpineol; 16. α-humulene; 17. neral; 18. α-terpineol; 19. valencene; 20. piperitone; 21. geraniol; 22. nerylacetate; 23. α-bisabolene; and 24. β-bisabolene.

second dimension column. After the heart-cutting, the remaining sample was backflushed at RT 12.20 min, and then the temperature program for ^{2}D GC started at RT 13.00 min. More than 10 compounds were clearly separated on the ^{2}D separation. Thanks to low dead volume MFD, EPC, and constant temperature of the transfer lines in the host GC oven, each delay time was very close to the theoretical value of 0.593 s, which is calculated using the following equation, and the deviation was less than 0.13 s (Sasamoto and Ochiai 2010):

$$F = [60\pi r^4/16\eta L]\left[(p_i^2 - p_o^2/p_o)(p_o/p_{ref})(T_{ref}/T)\right]$$

where F is outlet flow (milliliter per minute at T_{ref} and p_{ref}); r is column inner radius (centimeter); L is column length (centimeter); p_i is inlet pressure (absolute; dynes per centimeter); p_o is outlet pressure (absolute; dynes per centimeter); p_{ref} is reference pressure, typically 1 atm; T is column (oven) temperature (K); T_{ref} is reference temperature, typically 298 K; and η is carrier gas viscosity at column temperature (poise).

Analysis of green note compound in whiskey by stir bar sorptive extraction–thermal desorption selectable one-dimensional or two-dimensional gas chromatography–mass spectrometry with simultaneous olfactometry

The SBSE-TD-^{1}D GC–O/MS was initially performed to screen the potent aroma components in a single malt whiskey sample. Figure 2.5 shows a

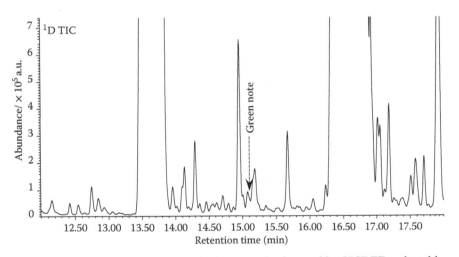

Figure 2.5 ^{1}D TIC of a single malt whiskey sample obtained by SBSE-TD–selectable ^{1}D/^{2}D-GC–O/MS, ^{1}D column: DB-Wax, 30 m × 0.25 mm i.d. × 0.25 μm d_f, 40°C–10°C/min–240°C.

TIC of single malt whiskey. A green note compound was detected with olfactometry at the RT 15.10 min. However, we could not identify this compound because this peak was coeluted with other peaks. Therefore, a heart-cut in the RT 14.96–15.24 min (17 s) region of the single malt whiskey was performed to transfer the green note compound to the second dimension. Figure 2.6 shows both the ¹D and the ²D TICs and the corresponding olfactometric signals of the single malt whiskey sample. One-dimensional GC–O/MS was done till 14.96 min, and the heart-cut region from 14.96 to 15.24 min was transferred to the second dimension.

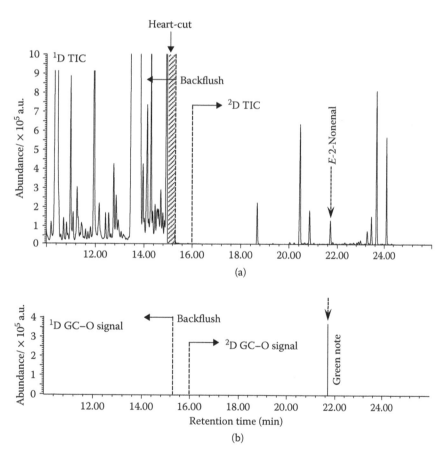

Figure 2.6 ¹D/²D TIC and the corresponding olfactometric signal of a single malt whiskey sample obtained by SBSE-TD–selectable ¹D/²D GC–O/MS. (a) ¹D/²D TIC. (b) ¹D/²D GC–O signal. ¹D column: DB-Wax, 30 m × 0.25 mm i.d. × 0.25 μm d_f, 40°C–10°C/min–240°C; ²D column: DB-5, 10 m × 0.18 mm i.d. × 0.40 μm d_f, 40°C–10°C/min–240°C.

The ^{2}D separation was performed from 16 min after pressure stabilization. The peak of interest was clearly separated from other compounds and was detected with olfactometry on the second dimension. A well-defined mass spectrum was obtained for the detected peak. From a National Institute of Standards and Technology (NIST) library search, the detected peak was identified as E-2-nonenal (Figure 2.7). The E-2-nonenal peak is known as an important odor component of aged beer (Santos et al. 2003; Ochiai et al. 2003). The odor description is "fatty, waxy, cucumber, cardboard, etc.," and odor threshold in fresh beer is as low as 0.1 ng·mL^{-1} (Meilgaard, Elizondo, and Moya 1970). Since the mass spectrum of E-2-nonennal is not so unique (e.g., typical aliphatic compound), LRI for the E-2-nonenal peak on the second dimension (DB-5) was calculated for positive identification. The calculated LRI of 1156 was compared with the average LRI from the literature, which was obtained from the Aroma Office ^{2}D database (Gerstel K. K./Nishikawa Keisoku Co., Ltd., Tokyo, Japan). The Aroma Office ^{2}D database includes 72,120 entries of LRI of aroma compounds and literature sources. The deviation between the average LRI of 1160 ($n = 30$) from the value from the literature and the calculated LRI was only 4 units. Thus, the green note compound

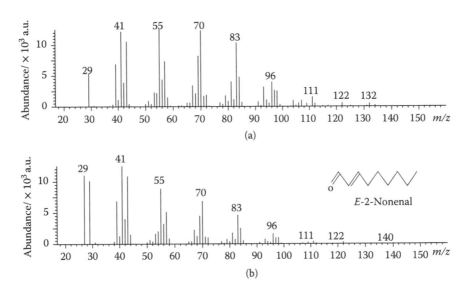

Figure 2.7 A NIST library search result for the green note peak on the ^{2}D separation. (a) Mass spectrum from the green note peak. (b) Mass spectrum of E-2-nonenal in the NIST library.

detected was assigned a probable identification as E-2-nonenal based on both mass spectral information and the ²D LRI. Finally, the E-2-nonenal standard was obtained and determination was carried out in six replicate analyses using a 4-points standard addition calibration between 10 and 60 ng·mL⁻¹. Good linearity was achieved with a correlation coefficient (r^2) of 0.9999. The E-2-nonenal was determined at a trace level of 11 ng·mL⁻¹ (relative standard deviation [RSD] = 1.7 %, n = 5). Wanikawa et al. (2002) reported that malt whiskies with high green note scores organoleptically contained more of E-2-nonenal; E; Z-2,6-nonadienal; and nonan-2-ol than those with low scores. They revealed five green note compounds including E-2-nonenal in malt whiskey by liquid–liquid extraction (LLE) in combination with GC–O/MS and ²D GC–O/MS. This type of study can be done with SBSE-TD in combination with the selectable ¹D/²D GC–O/MS system based on a single GC–MS system without any change in instrument setup.

Analysis of sulfur compounds in whiskey by large volume full evaporation technique–thermal desorption–selectable one-dimensional or two-dimensional gas chromatography–mass spectrometry with simultaneous element-specific detection

The LVFET-TD-¹D GC–NPD/PFPD/MS was initially performed to screen the sulfur compounds in a single malt whiskey sample. Figure 2.8 shows the ¹D TIC, the corresponding ¹D NPD chromatogram, and the corresponding ¹D PFPD chromatogram (S-trace) of the single malt whiskey sample. Although the NPD and the PFPD pinpointed numerous nitrogen (or phosphorus) compounds and sulfur compounds, coeluting sample matrix significantly interfered in extracting mass spectra of those compounds. For example, in the RT 17–18 min region although five sulfur peaks were clearly detected on the PFPD chromatogram, these peaks were completely buried in the TIC (Figure 2.9). Therefore, a heart-cut in the RT 17–18 min (60 s) region was performed to transfer the selected sulfur peaks and coeluting nitrogen compounds and sample matrix to the cryotrap. Heating the cryotrap to start the ²D GC separation gave a good separation of sulfur compounds and nitrogen (or phosphorus) compounds (Figures 2.10 and 2.11).

One sulfur compound and one sulfur/nitrogen compound were identified as 2-formylthiophene and 2-acetylthiazole based not only on mass spectral and elemental information but also on the ²D LRI (Figure 2.12, Table 2.2). Mac Namara (1993) indicated that the identification of

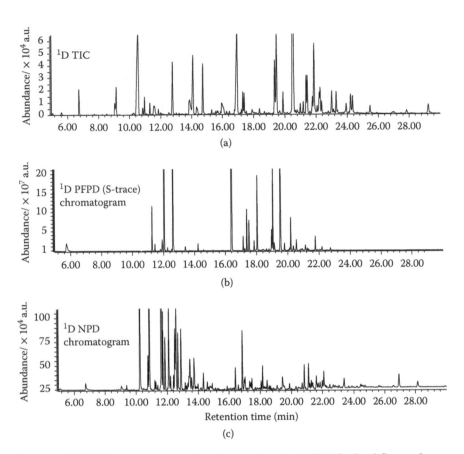

Figure 2.8 ^{1}D TIC and nitrogen phosphorus detector (NPD)/pulsed flame photo-metric detector (PFPD) chromatograms of a single malt whiskey sample obtained by SBSE-TD–selectable ^{1}D/^{2}D GC–NPD/PFPD/MS. (a) ^{1}D TIC. (b) ^{1}D PFPD chromatogram. (c) ^{1}D NPD chromatogram. DB-Wax; ^{1}D column: 30 m × 0.25 mm i.d. × 0.25 μm d_f; 40°C–10°C/min–240°C.

medium-volatile sulfur compounds (usually present at the sub-microgram per milliliter level) in whiskey is difficult and generally requires a non-routine analytical strategy. He revealed 17 sulfur compounds including 2-formylthiophene and 2-acetylthiazole from the starting sample of 2 L of whiskey by the combination of ^{2}D preparative GC, ^{2}D GC–SCD, and ^{2}D GC–MS. Thanks to the combined approach with LVFET-TD and select-able ^{1}D/^{2}D GC–NPD/PFPD/MS, hydrophilic sulfur compounds such as 2-formylthiophene (log K_{ow}: 1.53) and 2-acetylthiazole (log K_{ow}: 0.67) could be positively identified from only a 100-μL sample volume of the single malt whiskey.

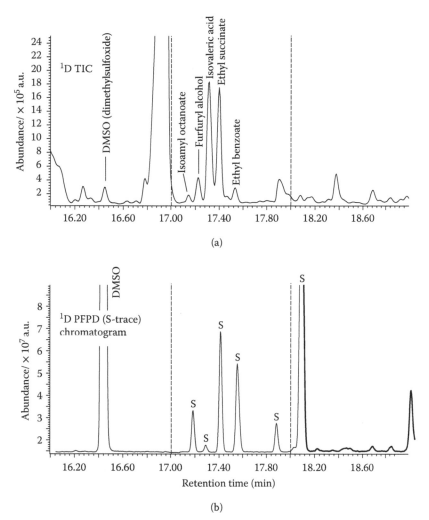

Figure 2.9 ¹D TIC and PFPD chromatogram of a single malt whiskey sample obtained by SBSE-TD–selectable ¹D/²D GC–NPD/PFPD/MS. (a) ¹D TIC. (b) ¹D PFPD chromatogram. S denotes unknown sulfur compound; DB-Wax, ¹D column: 30 m × 0.25 mm i.d. × 0.25 μm d_f, 40°C–10°C/min–240°C.

Selectable one-dimensional or two-dimensional gas chromatography–olfactometry/mass spectrometry with preparative fraction collection

In certain cases, ¹D/²D GC–O/MS is not able to produce high-quality mass spectra for the compounds detected by olfactory organs on ²D separation (no peaks on the ²D TIC at the corresponding RTs),

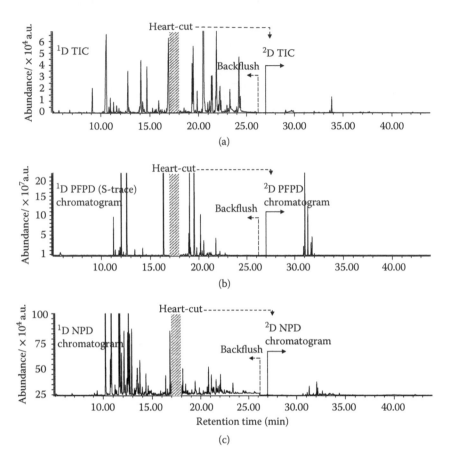

Figure 2.10 ^{1}D/^{2}D TIC and NPD/PFPD chromatograms of a single malt whiskey sample obtained by SBSE-TD–selectable ^{1}D/^{2}D GC–NPD/PFPD/MS. (a) ^{1}D/^{2}D TIC. (b) ^{1}D/^{2}D PFPD chromatogram. (c) ^{1}D/^{2}D NPD chromatogram. ^{1}D column: DB-Wax, 30 m × 0.25 mm i.d. × 0.25 μm d_f, 40°C–10°C/min–240°C; ^{2}D column: DB-5, 10 m × 0.18 mm i.d. × 0.40 μm d_f, 40°C–10°C/min–240°C.

particularly when analyzing highly complex aromas. In this case, it is essential to have an enrichment step before final MS detection. Ochiai and Sasamoto (in press) proposed an enrichment method for ^{1}D/^{2}D GC–O/MS using a PFC system (^{1}D/^{2}D GC–O/MS with PFC). The main advantages of this system are the simple and fast selection of ^{1}D GC–O/ MS, ^{2}D GC–O/MS, ^{1}D GC–PFC, or ^{2}D GC–PFC operation with a mouse

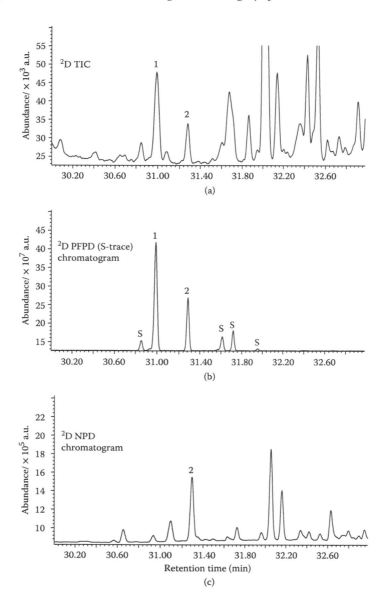

Figure 2.11 ¹D/²D TIC and NPD/PFPD chromatograms of a single malt whiskey sample obtained by SBSE-TD–selectable ¹D/²D GC–NPD/PFPD/MS. (a) ²D TIC. (b) ²D PFPD chromatogram. (c) ²D NPD chromatogram. 1. 2-formylthiophene; 2. 2-acetylthiazole; S. unknown sulfur compound; ¹D column: DB-Wax, 30 m × 0.25 mm i.d. × 0.25 μm d_f, 40°C–10°C/min–240°C; ²D column: DB-5, 10 m × 0.18 mm i.d. × 0.40 μm d_f, 40°C–10°C/min–240°C.

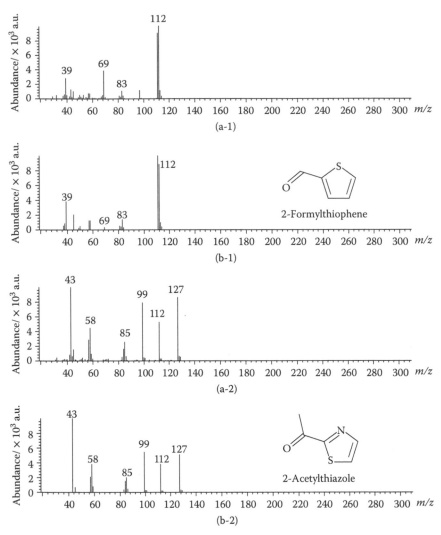

Figure 2.12 Measured mass spectra of 2-formylthiophene (a-1) and 2-acetylthia-zole (a-2) in a sigle malt whiskey sample obtained by large volume full evapora-tion technique–TD ¹D/²D GC–NPD/PFPD/MS, and Wiley library mass spectra of 2-formylthiophene (b-1) and 2-acetylthiazole (b-2).

click (without any change in the instrumental setup) and the total transfer of enriched compounds with TD on the same system for iden-tification with ²D GC–O/MS analysis. Figure 2.13 shows the schematic flow diagram for ¹D/²D GC–O/MS with PFC. The system consists of, in addition to the components of ¹D/²D GC–O/MS, a second Deans switch

Table 2.2 Calculated LRI on ²D GC and Average LRI from the Literature

			²D TIC			
	RT (min)	Calculated LRI	Average of LRIs in the literature[a]	Minimum LRI in the literature	Maximum LRI in the literature	Deviation[b]
2-Formylthiophene (isomer)	30.99	995	1002 ($n = 4$)	1000	1010	7
2-Acetylthiazole	31.28	1013	1021 ($n = 6$)	1018	1025	8

[a] The average LRIs were obtained from Aroma Office ²D database (Gerstel K. K./Nishikawa Keisoku Co., Ltd., Tokyo, Japan).
[b] Deviation was the difference between calculated LRI and average LRI.

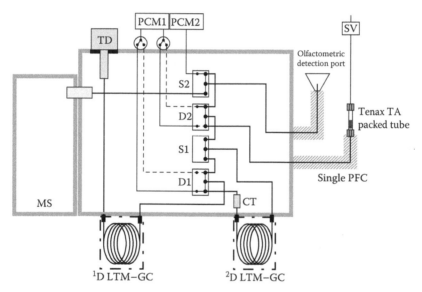

Figure 2.13 Schematic flow diagram of a selectable ¹D/²D GC–O/MS with PFC system. TD: thermal desorption inlet; PCM1 and PCM2: pressure control modules; D1 and D2: Deans switches; S1 and S2: splitters; and SV: solenoid valve.

(Deans 2), the PCM 2, and the single PFC module. Both the outlets of ¹D and ²D columns were merged and then connected to Deans 2 instead of a splitter, which was in turn connected to a mass spectrometer and an ODP. The outlet flow is switched to the splitter or the single PFC module by Deans 2, which is controlled by PCM 2. The system can provide not only ¹D or ²D GC–O/MS analysis but also PFC on both ¹D and ²D GC separations.

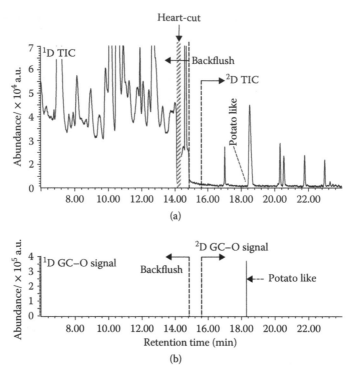

Figure 2.14 ^{1}D/^{2}D TIC and the corresponding olfactometric signal of a soup stock sample obtained by SBSE-TD–selectable ^{1}D/^{2}D GC–O/MS. (a) ^{1}D/^{2}D TIC. (b) ^{1}D/^{2}D GC–O signal. ^{1}D column: DB-Wax, 10 m × 0.18 mm i.d. × 0.18 μm d_f, 40°C–10°C/min–240°C. ^{2}D column: DB-5, 10 m × 0.18 mm i.d. × 0.40 μm d_f, 40°C–10°C/min–240°C.

Figure 2.14 shows both ^{1}D and ^{2}D TICs, and olfactometric signals of the soup stock obtained by SBSE-TD-^{1}D/^{2}D GC–O/MS. A potato-like compound was clearly detected with olfactometry on the ^{2}D separation. However, no detectable peak was found on the ^{2}D TIC because of a lack of detectability in scan mode after a single SBSE. After PFC enrichment with 12 injection cycles (enrichment of the potato-like compound from 12 stir bars) and ^{2}D separation, although the peak of interest was still almost buried in the ^{2}D TIC, specific mass chromatograms (e.g., *m/z* values of 48, 61, 76, and 104) were found at the time corresponding to the olfactometric signal of the potato-like compound (Figure 2.15). From a NIST library search, the detected peak was identified as methional (Figure 2.16). The odor description of methional is "boiled and baked potato," and odor threshold is 0.2 ppb (Rowe 2000). The LRI for the

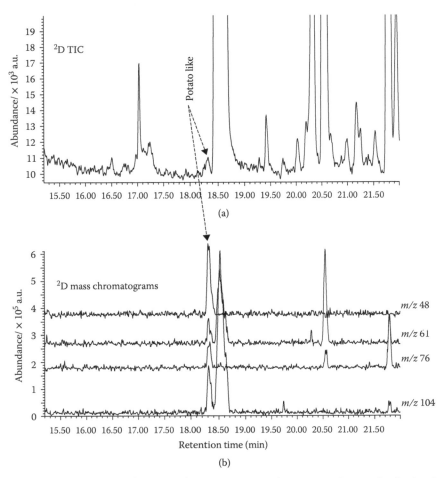

Figure 2.15 ²D TIC and ²D mass chromatograms of a soup stock sample obtained by SBSE–TD–selectable ¹D/²D GC–O/MS. (a) ²D TIC. (b) ²D mass chromatograms (*m/z* values of 48, 61, 76, and 104). ¹D column: DB-Wax, 10 m × 0.18 mm i.d. × 0.18 μm d_f, 40°C–10°C/min–240°C; ²D column: DB-5, 10 m × 0.18 mm i.d. × 0.40 μm d_f, 40°C–10°C/min–240°C.

methional peak on the second dimension (DB-5) was calculated for positive identification. The calculated LRI of 893 was compared with the average LRI from the literature, which was obtained from the Aroma Office ²D database (Gerstel K. K./Nishikawa Keisoku). The deviation between the average LRI of 901 (*n* = 15) from the literature and the calculated LRI was 8 units.

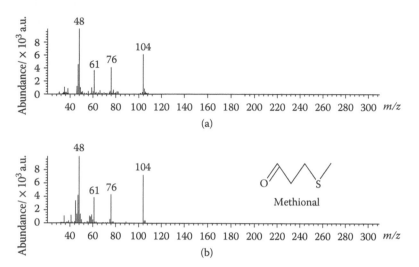

Figure 2.16 Measured mass spectra of methional (a) in a soup stock sample obtained by SBSE–preparative fraction collection enrichment with 12 injection cycles and subsequent TD ^{1}D/^{2}D GC–O/MS analysis, and NIST library mass spectra of methional (b).

Conclusion

Selectable ^{1}D/^{2}D GC–MS is a novel MDGC–MS system used for the analysis of trace amounts of odor compounds in complex samples. This system can perform ^{1}D GC–MS and ^{2}D GC–MS analyses (with a mouse click) without any change in the instrumental setup and can provide the unique capabilities of simultaneous mass spectrometric and olfactometry or element-specific detection in both ^{1}D and ^{2}D GC under constant split ratio. Coupled with PFC using an adsorbent-packed tube and subsequent TD, ^{1}D/^{2}D GC–MS allows identification of ultratrace amounts of odor compounds in complex samples, with reasonable analysis times.

Acknowledgments

The author thanks Kikuo Sasamoto of Gerstel K. K. and Edward Pfannkoch of Gerstel, Inc., for their kind support in preparing this chapter.

References

Adahchour, M., J. Been, R. J. J. Vreuls, and U. A. Th. Brinkman. 2006. Recent developments in comprehensive two-dimensional gas chromatography (GC × GC). *Trends Anal Chem* 25(5):438–54.

Bertsch, W. 1990. Multidimensional gas chromatography. In *Multidimensional Chromatography*, ed. H. Cortes, 74–144. New York: Marcel Dekker.

Blank, I. 2002. Gas chromatography-olfactometry in food aroma analysis. In *Flavor, Fragrance, and Odor Analysis*, ed. R. Marsili, 297–331. New York: Marcel Dekker.

David, F., and M. S. Klee. 2009. Analysis of suspected flavor and fragrance allergens in perfumes using two-dimensional GC with independent column temperature control using an LTM oven module, Agilent Technologies publication 5990-3576N, February 2009. http://www.chem.agilent.com/Library/applications/5990-3576EN.pdf, accessed August 10, 2011.

David, F., and P. Sandra. 2007. Stir bar sorptive extraction for trace analysis. *J Chromatogr A* 1152:54–69.

Deans, D. R. 1968. A new technique for heart cutting in gas chromatography. *Chromatographia* 1:18–22.

Feyerhem, F., R. Lowe, J. Stuff, and D. Singer. 2007. Rapid multidimensional GC analysis of trace drugs in complex matrices, GERSTEL AppNote 8/2007. http://www.gerstel.com/pdf/p-gc-an-2007-08.pdf, accessed August 10, 2011.

Hoffmann, A., O. Lerch, and V. Hudewenz. 2009. Fragrance profiling of consumer products using a fully automated dynamic headspace system, GERSTEL AppNote 8/2009. http://www.gerstel.com/pdf/p-gc-an-2009-08.pdf, accessed August 10, 2011.

Luong, J., R. Gras, G. Yang, L. Sieben, and H. Cortes. 2007. Capillary flow technology with multi-dimensional gas chromatography for trace analysis of oxygenated compounds in complex hydrocarbon matrices. *J Chromatogr Sci* 45:664–70.

Luong, J., R. Gras, G. Young, H. Cortes, and R. Mustacich. 2008. Multidimensional gas chromatography with capillary flow technology and LTM-GC. *J Sep Sci* 31:3385–94.

Mac Namara, K. 1993. Investigation of medium volatile sulfur compounds in whiskey. In *Elaboration et Connaissance des Spiritueux*, ed. R. Cantagrel, 385–391. Paris: Lavoisier & Doc.

Markelov, M., and O. A. Bershevits. 2001. Methodologies of quantitative headspace analysis using vapor phase sweeping. *Anal Chim Acta* 432:213–27.

Markelov, M., and J. P. Guzowski. 1993. Matrix independent headspace gas chromatographic analysis. The full evaporation technique. *Anal Chim Acta* 276:235–45.

Meilgaard, M., A. Elizondo, and E. Moya. 1970. Study of carbonyl compounds in beer, part 2. Flavor and flavor thresholds of aldehydes and ketones added to beer. *MBAA Tech Q* 7:143–9.

Nitz, S., H. Kollmannsberger, and F. Drawert. 1989. Determination of sensorial active trace compounds by multi-dimensional gas chromatography combined with different enrichment techniques. *J Chromatogr* 471:173–85.

Ochiai, N., and K. Sasamoto. (In press). Selectable one-dimensional or two-dimensional gas chromatography–olfactometry/mass spectrometry with preparative fraction collection for analysis of ultra-trace amounts of odor compounds. *J Chromatogr A* 2011 1218:3180–5.

Ochiai, N., K. Sasamoto, S. Daishima, A. C. Heiden, and A. Hoffmann. 2003. Determination of stale-flavor carbonyl compounds in beer by stir bar sorptive extraction with in-situ derivatization and thermal desorption–gas chromatography–mass spectrometry. *J Chromatogr A* 986:101–10.

Pfannkoch, E. A., and J. A. Whitecavage. 2005. A selectable single or multidimensional GC system with heart-cut fraction collection and dual detection for trace analysis of complex samples, GERSTEL AppNote 4/2005. http://www.gerstel.com/pdf/p-gc-an-2005-04.pdf, accessed August 10, 2011.

Rowe, D. 2000. More fizz for your buck: High impact aroma chemicals. *Perfumer Flavorist Mag* 25(5):1–19.

Santos, J. R., J. R. Carneiro, L. F. Guido, P. J. Almedia, J. A. Rodrigues, and A. A. Barros. 2003. Determination of E-2-nonenal by high-performance liquid chromatography with UV detection: Assay for the evaluation of beer ageing. *J Chromatogr A* 985:395–402.

Sasamoto, K., and N. Ochiai. 2008. Heart-cutting 2D GC–MS with micro-fluidic Deans switch and low thermal mass (LTM) GC, Paper presented at the 32nd International Symposium on Capillary Chromatography, Riva del Garda.

Sasamoto, K., and N. Ochiai. 2010. Selectable one-dimensional or two-dimensional gas chromatography—mass spectrometry with simultaneous olfactometry or element-specific detection. *J Chromatogr A* 1217:2903–10.

Wanikawa, A., K. Hosoi, T. Kato, and K. Nakagawa. 2002. Identification of green note compounds in malt whiskey using multidimensional gas chromatography. *Flavour Fragr J* 17:207–11.

Zellner, B., A. Casilli, P. Dugob, G. Dugoa, and L. Mondello. 2007. Odour fingerprint acquisition by means of comprehensive two-dimensional gas chromatography-olfactometry and comprehensive two-dimensional gas chromatography/mass spectrometry. *J Chromatogr A* 1141:279–86.

chapter three

Simple derivatization prior to stir bar sorptive extraction to improve extraction efficiency and chromatography of hydrophilic analytes

Acylation of phenolic compounds

Ray Marsili

Contents

As building blocks of plants and in a wide range of industrial applications, phenolic compounds are commonly found in water and food products. Because of their toxicity and unpleasant organoleptic properties, some phenols have been included in the priority pollutants list of the U.S. Environmental Protection Agency (EPA).

Analysis of phenols can be challenging. Since phenols are polar analytes, they tend to be water soluble and are difficult to extract with organic solvents. In fact, small polar analytes are often the most difficult analytes to extract. Therefore, classical liquid–liquid extraction is not well suited as a preconcentration technique prior to chromatographic analysis. Solid-phase extraction (SPE) of phenols as an enrichment step can be problematic because of their high polarity.

Although conversion of phenols to lower polarity phenolic derivatives adds analytical steps to a procedure, the investment in derivatization is worthwhile, resulting in sharper analyte peaks and improved detection limits.

Although liquid chromatography (LC) can be performed, gas chromatography (GC) is usually preferred because of its superior separation efficiency and high sensitivity, and the identification capability of today's modern GC mass spectrometer (MS) detectors. Although polar phenolic compounds tend to adsorb onto active sites in GC injectors and GC columns, resulting in broad and tailing peaks, this can be obviated by converting the phenolic compounds to less-polar derivatives.

Several methods have been described for preparing phenolic derivatives prior to GC analysis, but perhaps the most straightforward method is in situ derivatization prior to extraction. For example, a 10-mL water sample can be added to a 20-mL GC vial. Potassium carbonate (0.5 g) and acetic anhydride (0.5 g) are then added. The phenols are converted to the phenolates as acylation takes place (Figure 3.1).

After derivatization, the sample can be analyzed by solid-phase microextraction (SPME) GC–MS (e.g., using a 100 μm polydimethylsiloxane [PDMS] fiber) [1], or, for even lower detection limits, by stir bar sorptive extraction (SBSE).

Baltussen developed a novel sorptive extraction technique using a stir bar coated with PDMS and applied the technique to the analysis of phenolic compounds [2]. The technique, called stir bar sorptive extraction, is similar to SPME based on PDMS sorption. SBSE, however, offers several advantages over SPME, including larger mass of PDMS sorbent, which results in higher recoveries and higher sample capacity.

Since SBSE is an equilibrium technique, the extraction of solutes from the aqueous phase into the PDMS phase is determined by the partitioning coefficient. Studies have correlated this partitioning coefficient with the octanol–water distribution constant ($K_{o/w}$) [3].

Polar compounds like phenols are not well suited to extraction by the apolar PDMS phase. Fortunately, the problem can be circumvented by using a simple derivatization method that can be performed in aqueous

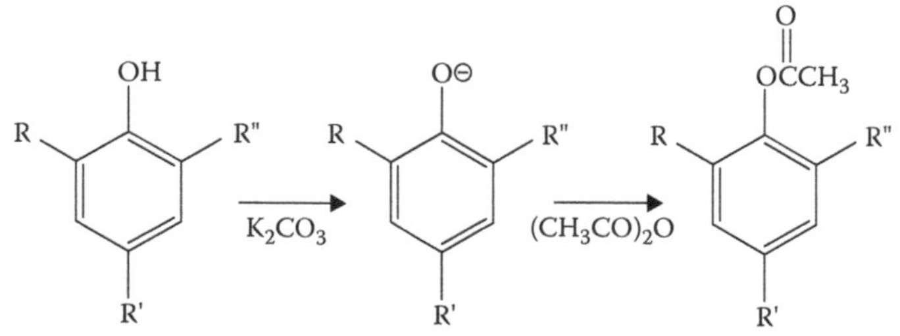

Figure 3.1 Conversion of phenols into phenolates as acylation takes place.

phase prior to SBSE extraction. Derivatization increases sample enrichment in the PDMS phase by converting the phenolics to phenolic acetates, which have higher octanol/water coefficients (log $K_{o/w}$) compared with the phenolic compounds from which they originate. Furthermore, this simple derivatization results in sharper peaks, decreases peak tailing, improves chromatographic analysis by increasing peak resolution, and, in some cases, shifts phenolic peak elution away from interfering chromatographic peaks.

This chapter discusses examples of how analytical sensitivity and recovery of phenolic compounds can be improved using aqueous acylation prior to SBSE GC–MS.

Determination of alkylphenols and bisphenol A in river water by stir bar sorptive extraction

Alkylphenols and bisphenol A (BPA) are considered to be endocrine disrupters. In order to accurately assess potential risks, highly sensitive methods are required. For example, the Japan Environmental Agency has proposed that these compounds be analyzed directly or as ethyl derivatives by gas chromatography–mass spectrometry (GC–MS) after 1000-fold concentration by SPE. According to the agency, BPA is determined as its trimethyl silyl derivative by GC–MS after SPE. LC–MS has also been applied to the study of these compounds.

BPA, a food contaminant that has been linked to a host of potential health risks in numerous studies, has been of significant concern to the food industry over the past decade. BPA, which mimics the hormone estrogen, is used in the manufacture of some plastics and has been linked to behavioral problems and heart ailments in children. Manufacturers and retailers of plastic baby bottles and water bottles have been switching to BPA-free plastics. Food and beverage can liners use BPA in their epoxy resins, but some companies have introduced BPA-free cans. Plastic food storage containers, dental sealants, polycarbonate plastics used in helmets, computer cases, and other products, as well as thermal paper cash register receipts [4] have all been shown to contain worrisome levels of BPA. In 1963, the FDA classified BPA as an indirect food additive and listed it as GRAS ("generally recognized as safe"). Canada recently became the first country to declare BPA a toxic substance.

Nakamura et al. [5] proposed a sensitive but simplified method for the determination of seven key alkylphenols and BPA by SBSE with in situ derivatization-thermal desorption (TD-GC–MS). For 4-*tert*-butylphenol and BPA, in situ acylation improved the responses in SBSE-TD-GC–MS compared to SBSE-TD-GC–MS without derivatization. The method detection limits ranged from 0.1 to 3.2 ng/L. The recoveries of the analytes from a river water sample spiked with standards at 10 and 100 ng/L were

85.3%–105.9% (RSD = 3.0%–11.0%) and 88.3%–105.8% (RSD = 1.6%–8.3%), respectively.

A simple acylation SBSE extraction procedure was used. Stir bars coated with PDMS (10 × 0.5 mm) were added to 10 mL of river water in a 10-mL glass vial. Potassium carbonate (0.5 g) and acetic acid anhydride (0.5 mL) were added, and a PTFE-faced septum cap was used for crimping. The water sample was stirred for 60 min at 1000 rpm at room temperature. The stir bar was then removed from the vial and briefly dried with a lint-free tissue and finally transferred to an empty glass thermal desorption vial for thermal desorption.

The comparison of the responses (peak area of total ion chromatograms in the scan mode) of SBSE sampling without derivatization at pH 3.5 and SBSE sampling with acylation is shown in Table 3.1. Detection limits for 4-*tert*-butylphenol and BPA were dramatically increased after acylation. Conversion of phenolic compounds to acetyl derivatives can improve not only chromatographic analysis but also sample enrichment in the PDMS phase because of an increase in log $K_{o/w}$ values. Figure 3.2 shows the SIM chromatograms obtained by TD-GC–MS after 60-min SBSE with in situ acylation of a natural water sample (10 mL) fortified at 50 ng/L. The method was able to detect alkylphenols and BPA in river water in nanograms per liter levels.

This technique by Nakamura et al. could be applied to more hydrophilic phenolic compounds (those with smaller alkyl groups) that have significant odor activity in food and water. Acylation would produce derivatives with higher log $K_{o/w}$ values and make them more amenable to SBSE.

Table 3.1 Comparison of Responses of SBSE Sampling without Derivatization and SBSE Sampling with Acetylation

Analyte	SBSE without derivatization	SBSE with derivatization
4-*tert*-butylphenol	1,899,038	22,328,804
4-*n*-pentylphenol	13,407,426	23,493,322
4-*n*-hexylphenol	24,872,745	27,807,443
4-*tert*-octylphenol	23,474,733	20,760,161
4-*n*-heptylphenol	33,787,968	30,229,474
Nonylphenol (isomers)	3,556,959	2,764,445[a]
4-*n*-octylphenol	22,308,813	12,196,731
Bisphenol A	368,853	38,954,272

Source: Data from Nakamura, S. and S. Daishima. 2004. *J Chromatography A* 1038:291–294.

SBSE = stir bar sorptive extraction.

[a] Responses of nonylphenol were calculated on the basis of a representative peak among its isomers.

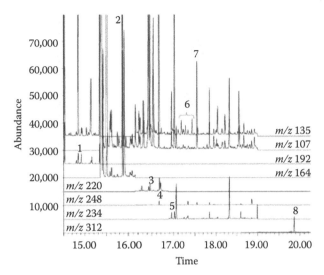

Figure 3.2 SIM chromatograms by SBSE with in situ acylation and TD-GC–MS of fortified natural water spiked at 50 ng/L. (1) 4-*tert*-butylphenol; (2) 4-*n*-pentyl phenol; (3) 4-*n*-hexylphenol; (4) 4-*tert*-octylphenol; (5) 4-*n*-heptylphenol; (6) nonylphenol isomers; (7) 4-*n*-octylphenol; (8) bisphenol A. (From Nakamura, S. and S. Daishima. 2004. *J. Chromatography A* 1038:291–294.)

Identification of smoke-taint chemicals in wine by SBSE GC–MS techniques

Unprecedented wildfires in northern California in 2008 caused considerable concerns in the winemaking industry. Chong at E&J Gallo Winery, Modesto, California, identified the volatile phenols guaiacol and 4-methyl guaiacol (aka *p*-creosol) as the primary smoke-derived markers in smoke-tainted wines [6]. Work by other researchers on smoke taint in wine supports the theory that guaiacol and *p*-creosol are significant contributors to the smoke-taint flavor in wine [7]. Other potential contributors to smoky aroma include 4-ethylguaiacol, 4-ethylphenol, eugenol, and furfural.

Sensory studies have characterized the malodor of smoke-tainted wine as smoky, dirty, burnt, and ash aroma [6,7]. Chong concluded that guaiacol and 4-methyl guaiacol were bound to glucose, arabinose, rhamnose, galactose, and xylose in smoke-exposed grapes. The glucose-bound compounds would likely be released by yeast glucosidase during fermentation and smoke-taint chemicals bound to other sugars would likely be released by acid hydrolysis during wine aging [8]. Chong developed an enzymatic hydrolysis technique to free bound smoke-taint chemicals, so they could be analyzed by headspace GC.

Our laboratory was asked to analyze a Cabernet Sauvignon wine sample that had been previously tested by another laboratory using enzymatic pretreatment followed by SPME GC–MS. Although sensory profiling clearly showed that the wine sample had a detectable smoke taint, SPME GC–MS failed to detect guaiacol or 4-methyl guaiacol. Our goal was to determine whether SBSE, a technique that usually shows higher sensitivity than SPME for many types of analytes, could be used to detect the smoke-taint chemicals in the wine sample. To maximize the recovery of smoke-taint analytes, which are polar phenolics, two SBSE methods were investigated: (1) sequential SBSE [9] and (2) acylation of wine prior to SBSE.

The methods were as follows:

Sequential SBSE:

1. 10 g of wine and 10 mL of water were added to a 50-mL Erlenmeyer flask.
2. A 2 cm × 0.5 mm Twister (47 µL) PDMS was added to the flask, and the diluted wine was stirred at 1000 rpm for 1.5 h.
3. 3 g of NaCl was added, and the sample was stirred using the same Twister for an additional 1.5 h.
4. The Twister was removed from the diluted wine sample, rinsed with distilled water, patted dry with a lintless cloth, and added to a thermal desorption tube.
5. The thermal desorption tube was thermally desorbed with the Gerstel TDU.

Acylation of wine followed by SBSE with salting out:

1. 10 g of wine plus 10 mL of water were added to a 50-mL Erlenmeyer flask.
2. 1 g of potassium carbonate was added, and the sample was stirred until the potassium carbonate dissolved (5 min) with a Teflon-coated micro-stir bar.
3. 1 mL of acetic anhydride was added, and the sample was stirred for 15 min (acylation step). Slow addition of acetic anhydride is recommended to prevent overflow of bubbling sample.
4. 3 g of NaCl was added, and the sample was stirred until NaCl dissolved. Then, the Teflon-coated micro-stir bar was removed.
5. A 2 cm × 0.5 mm Twister (47 µL PDMS) was added, and the sample was stirred for 3 h at 1000 rpm.
6. The Twister is removed from the diluted wine sample, rinsed with distilled water, patted dry with a lintless cloth, and added to a thermal desorption tube.
7. The thermal desorption tube was thermally desorbed with the Gerstel TDU.

An Rtx-50MS (Restek) was used, 30 m × 250 µm with a nominal film thickness of 0.25 µm. Constant flow at 1.5 mL/min (helium) was used. Volatiles trapped on the Gerstel Twister were desorbed in a Gerstel TDU. Volatiles were trapped at –100°C on a Gerstel CIS glass wool liner during desorption in a GERSTE TDU (at 280°C for 4 min).

Sequential SBSE detected 4-methyl guaiacol (*p*-creosol), which coeluted with 2-phenyl acetate. No guaiacol was detected by sequential SBSE. The sequential SBSE chromatogram and a magnified view of the region of chromatogram where *p*-creosol and 2-phenyl acetate are coeluting are shown in Figure 3.3a (TIC). The mass spectrum of the coeluted peaks is shown in Figure 3.3b.

While the sequential SBSE technique was encouraging because it was able to detect *p*-creosol, which SPME failed to do, it was disappointing because the *p*-creosol coeluted with another chemical and no guaiacol was detected. Acylation SBSE provided better results because the *p*-creosol acetate peak did not coelute with other chemicals, and guaiacol was easily detected.

Figure 3.4a shows the *p*-creosol acetate peak at extracted ion *m/z* 138. Figure 3.4b shows the mass spectrum of the *p*-creosol peak in the wine compared to a NIST library mass spectrum for *p*-creosol.

Figure 3.5a shows guaiacol acetate at extracted ion *m/z* 124 by SBSE after acylation. The retention time is 16.15 minutes. Figure 3.5b shows that the guaiacol acetate peak in the wine sample closely matches the spectrum for guaiacol acetate in the NIST library.

In conclusion, SBSE techniques are able to detect smoke-taint chemicals in wine where SPME failed to do so. Sample preparation using acylation to convert guaiacol and *p*-creosol to their acetate forms followed by SBSE is even more effective for measuring smoke-taint chemicals (Figure 3.6). The acylated guaiacols have higher log $K_{o/w}$ values compared to their nonacetylated forms, and, therefore, are extracted with higher recovery. Furthermore, the acylation of *p*-creosol resulted in the separation of this chemical from a coeluting peak (2-phenylacetate) observed in the SBSE chromatogram. The acylation technique is relatively simple to perform and adds little sample preparation time.

Determination of cresols in casein by SBSE GC–MS: Maximizing SBSE recovery with acylation

p-Cresol has been reported as a bacterial metabolites of tyrosine. Researchers have linked phenolic compounds to barny/cow-like odors, specifically in milk proteins.

Caseins are widely used ingredients in food application because of their flavor stability and functional properties. Casein ingredients are used in cheese analogs, bakery, meat, confectionery products, nondairy

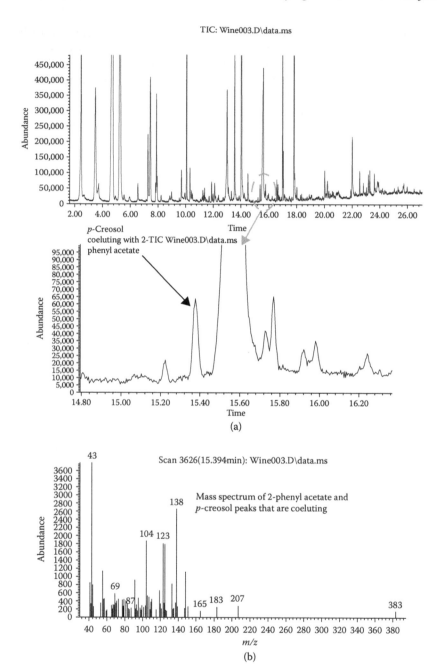

Figure 3.3 (a) SBSE chromatogram of smoke-tainted wine sample showing *p*-creosol coeluting with 2-phenyl acetate and no peak detected for guaiacol (prior to acylation). (b) Mass spectrum of peak at 15.394 min in previous chromatogram (Figure 3.3a) where *p*-cresol and 2-phenyl acetate coeluted.

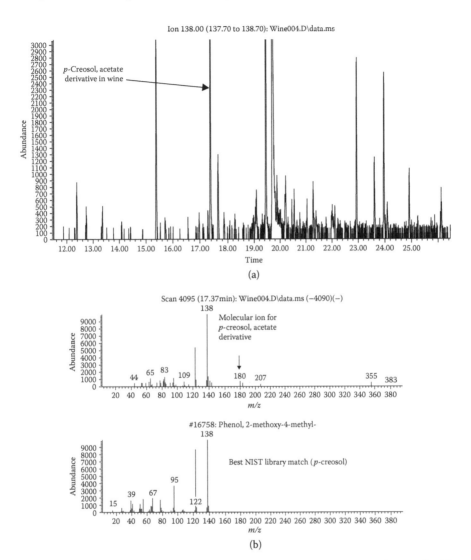

Figure 3.4 (a) SBSE chromatogram of smoke-tainted wine at extracted ion *m/z* 138 after acylation showing *p*-cresol peak with no coelution problem. (b) Comparison of *p*-creosol acetate peak in wine sample after acylation to *p*-creosol NIST library match.

creamers, and desserts. Casein, along with skim milk powder and ster-ilized concentrated milk, commonly suffers from a "stale" flavor, some-times characterized as "glue-like," "burnt feathers," or "wet dog-like." *o*-Aminoacetophenone was identified as a malodor chemical of stored skim milk powder [10,11]. Karagul-Yuceer et al. identified hexanoic acid, indole, guaiacol, and *p*-cresol as the major contributors to the typical ani-mal/wet dog-like malodor of rennet casein [12].

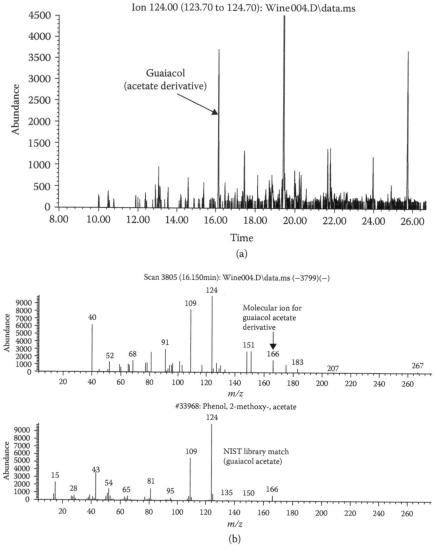

Figure 3.5 (a) Guaiacol acetate by SBSE after acylation. RT = 16.15 min; MW 166; extracted ion at *m/z* 124. (b) Guaiacol acetate by SBSE after acylation. RT = 16.15 min; MW 166; Guaiacol acetate peak in wine compared to guaiacol acetate in NIST library.

Since the odor threshold for *p*-cresol is approximately 50 ppb [13], an effective analytical method for monitoring cresols should demonstrate a sensitivity of 50 ppb. SBSE after acylation was found to be an appropriate analytical technique for the accurate determination of ppb levels of cresol isomers in casein powder.

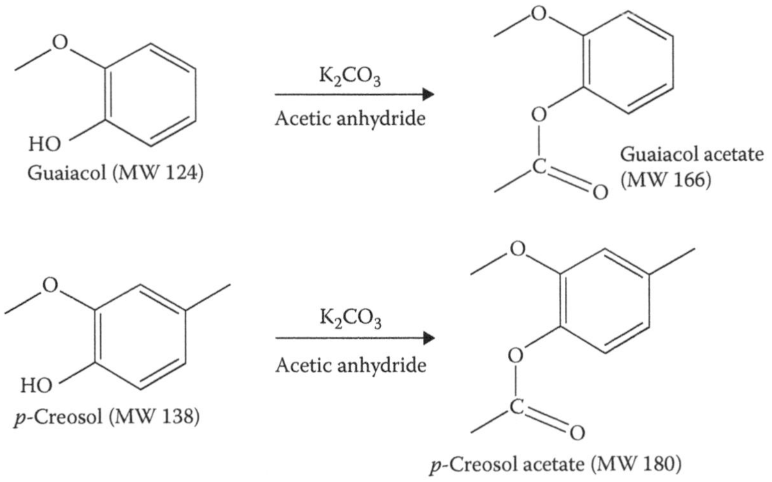

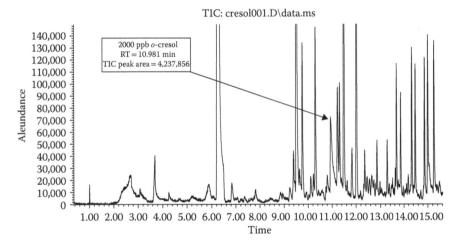

Figure 3.6 Derivatization/acylation reaction of guaiacols with potassium carbonate and acetic acid.

TIC: cresol001.D\data.ms

Figure 3.7 TIC of 2000 ppb cresol added to acid casein, demonstrating poor peak shape and inadequate sensitivity for *o*-cresol. 1-g casein sample treated with methanol and 6 M HCl prior to SBSE.

SBSE pretreatment of casein with methanol and 6 M HCl is employed to free the cresols that are bound to the casein. Initial testing showed that SBSE was not sensitive enough to detect 50 ppb cresol isomers in casein. For example, 1-g sample of casein in 10 mL of water was spiked with 2000 ppb *o*-cresol and treated with methanol and HCl and extracted with a 1 cm × 0.5 mm Twister. The resulting chromatogram is shown in Figure 3.7. The TIC of this sample demonstrates poor peak shape and inadequate sensitivity for *o*-cresol.

The log $K_{o/w}$ of *o*-cresol is 1.98. Because the cresol isomers are too hydrophilic to be extracted with high recovery from aqueous/methanol solutions, steps were devised to enhance the extraction recovery of cresols by SBSE. The following steps were used to maximize extraction recoveries of cresols from methanolic extracts of casein powder by SBSE; all the steps were important in achieving low ppb detection levels.

1. Salting out.
2. Most of the methanol from extracts was evaporated prior to SBSE.
3. Cresols were derivatized with acetic anhydride to increase their log $K_{o/w}$ values. Cresols were converted to their corresponding methylphenyl acetic acid esters.
4. The Agilent MS was operated in SIM instead of the Scan (TIC) mode.
5. Using a 2 cm × 0.5 mm Twister instead of a 1cm × 0.5 mm Twister increases sensitivity and is more appropriate when extracting large volumes of samples.
6. Splitless injection was used.

The mass spectra of *o*-cresol (Cas No. 95-48-7) and *o*-cresyl acetate (Cas No. 533-18-6) are shown in Figure 3.8.

The final sample preparation method includes the following steps:

1. Add 1.000 g of casein or standard mix solution plus 20 µL 6 M HCl to 3 mL water in a 40-mL GC vial. Stir vigorously and sonicate for 10 min to hydrate the casein.
2. Spike with 10 µL of deuterated *o*-cresol internal standard solution (0.04 µg/µL).
3. Add 6 mL of methanol and stir and sonicate for 10 min (to free-bound cresols from casein).

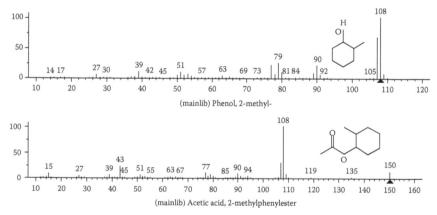

Figure 3.8 The mass spectra of *o*-cresol (Cas No. 95-48-7) and *o*-cresyl acetate (Cas No. 533-18-6).

4. Filter through a medium filter paper, washing filtrate with 3 mL of water. Evaporate off MeOH with gentle nitrogen stream and mild (55°C) heat.
5. Transfer to a 50-mL Erlenmeyer flask with 30 mL of distilled water.
6. Add 1 g of potassium carbonate and stir until dissolved.
7. Add 1 mL of acetic anhydride and stir for 15 min (acylation step).
8. Add 3 g of NaCl and stir until dissolved (salting out).
9. Add a 2 cm × 0.5 mm Twister and stir for 3 h at 1000 rpm.
10. Analyze in SIM at m/z 108, 115, and 150.

Note that a 30 m × 0.32 mm (df = 1 μm) DB-5 column was used because it was recommended in the literature for acetylated phenols. It is important to analyze reagent blank samples to screen reagents for cresol contamination.

Chromatograms of individual acylated cresols and deuterated o-cresyl acetate were analyzed (extracted ion m/z 108) to establish purity of standards

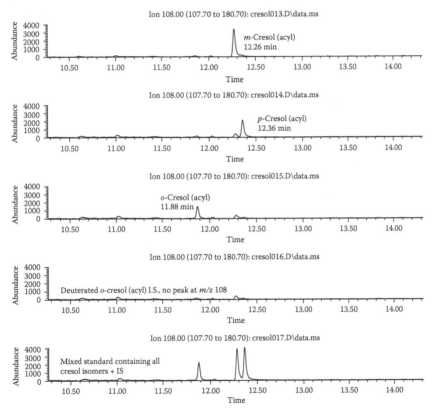

Figure 3.9 Chromatograms of individual acetylated cresols and deuterated o-cresyl acetate (internal standard) were analyzed (extracted ion m/z 108) to establish purity of standards and internal standard and to determine retention times. Each cresol concentration was 1 ppm spiked in control casein with no detectable cresols.

and internal standard and to determine retention times. Chromatograms are shown in Figure 3.9. Chromatograms of individual cresols and deuterated *o*-cresyl acetate internal standard were also examined at extracted ion *m/z* 115 in Figure 3.10. The deuterated *o*-cresyl acetate internal standard was added to the last two sample chromatograms; the internal standard peaks are circled.

Acylation significantly improves detection limits. This is dramatically illustrated in the chromatograms in Figure 3.11 (2000 ppb mixed cresol standard in scan mode at extracted ion of *m/z* 108).

Examples of significant experimental parameters, including odor thresholds for cresol isomers, typical concentrations of cresols in malodorous complaint caseins, and linear least squares correlation coefficients for standard curves for cresol isomers spiked in casein, are presented in Table 3.2.

In conclusion, the derivatization technique applied to polar analytes prior to SBSE expands the application potential of SBSE to phenols and many other flavor-important hydrophilic analytes, for example, amino acids and imidazoles. Acylation and other derivatization techniques should be investigated for these applications.

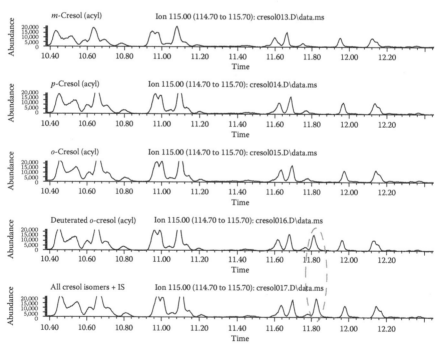

Figure 3.10 Chromatograms of individual acylated cresols and deuterated *o*-cresyl acetate (internal standard added to the last two samples) were analyzed (extracted ion *m/z* 115) showing internal standard peak (shown by ⬭). Each cresol concentration was 1 ppm spiked in control casein with no detectable cresols.

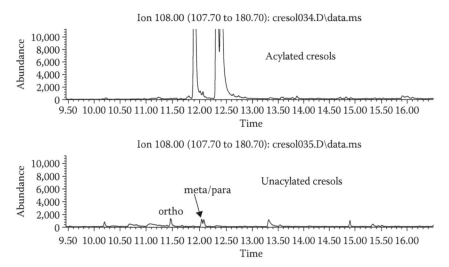

Figure 3.11 Chromatograms of *o*-, *m*-, and *p*-cresol comparing acylated cresols to nonacylated cresols. Extracted ion at *m/z* 108. 2000 ppb mixed cresol standard in scan mode at extracted ion of *m/z* 108.

Table 3.2 Experimental Parameters

Parameter	*o*-Cresol	*m*-Cresol	*p*-Cresol
Odor threshold[a]	650 ppb [14,15]	680 ppb [14,15]	55 ppb [13]
Typical concentrations in malodorous rennet caseins (nine samples tested)[b,c]	24.2 ppb ± 4.9%	75.1 ppb ± 4.9%	32.3 ppb ± 9.5%
Linear least squares correlation coefficients for standard calibration curves[c,d]	0.9982	0.9908	0.9979

[a] Literature values. (See references.)
[b] Nine malodor complaint rennet caseins. Average value ± percent standard deviation.
[c] Analytical testing method based on SBSE after acylation and SIM GC–MS, as described in chapter as "the final sample preparation method."
[d] Five standard solutions of 4, 40, 100, 200, and 500 ppd spiked in control casein with no cresols detected and no odor complaint.

References

1. Shirey, R. 1999. SPME fibers and applications, in solid phase microextraction: A practical guide. In *Varian Chromatography Systems*, ed. S. A. Scheppers Wercinski, 78. Walnut Creek, CA, Marcel Dekker, Inc.
2. Baltussen, H. A. 2000. *New Concepts in Sorption Based Sample Preparation for Chromatography, 137-144*. Eindhoven: Technische Universiteit Eindhoven.

3. Baltussen, H. A., P. Sandra, F. David, and C. Cramers. 1999. Stir bar sorptive extraction (SBSE), a novel extraction technique for aqueous samples. Theory and principles. *J Microcolumn Sep* 24:737–747.
4. Mendum, T., E. Stoler, H. VanBenschoten, and J. C. Warner. 2010. Concentrations of bisphenol A in thermal paper. *Green Chem Lett Rev* 28:1–6.
5. Nakamura, S. and S. Daishima. 2004. Simultaneous determination of alkyl-phenols and bisphenol A in river water by stir bar sorptive extraction with in-situ acylation and thermal desorption GC–MS. *J Chromatography A* 1038:291–294.
6. Chong, H. H. Smoke taint aroma assessment in 2008 California grape harvest. *Cornucopia*. AGFG abstracts for the 240th American Chemical Society National Meeting. August 22, 2010, Boston.
7. Kennison, K., K. L. Wilkinson, H. G. Williams, J. H. Smith, and M. R. Gibberd. 2007. Smoke-derived taint in wine: Effect of postharvest smoke of grapes on the chemical composition and sensory characteristics of wine. *J Ag Food Chem* 55:10897–10901.
8. Kennison, K., M. R. Gibberd, A. P. Polinitz, and K. L. Wilkinson. 2008. Smoke-derived taint in wine: The release of smoke-derived volatile phenols during fermentation of Merlot juice following grapevine exposure to smoke. *J Ag Food Chem* 56:7379–7383.
9. Ochiai, N., K. Sasamoto, J. Kanda, and E. Pfannkoch. 2008. Sequential stir bar sorptive extraction for uniform enrichment of trace amounts of organic pollutants in water samples. *J Chrom A* 1200:72–79.
10. Parks, O. W., D. P. Schwartz, and M. Keeney. 1964. Identification of o-amino-acetophenone as a flavor compound in stale dry milk. *Nature* 202:185.
11. Karagul-Yuceer, Y., M. A. Drake, and K. R. Cadwallader. 2001. Aroma-active components of nonfat dry milk. *J Agric Food Chem* 49:2948.
12. Karagul-Yuceer, Y., K. N. Vlahovich, M. A. Drake, and K. R. Cadwallader. 2003. Characteristic aroma components of rennet casein. *J Agric Food Chem* 51:6797.
13. Buttery, B. G., J. G. Turnbaugh, and L. C. Ling. 1988. Contribution of volatiles to rice aroma. *J Agric Food Chem* 36:1006.
14. Fazzalari, F. A., ed. 1978. *Compilation of odor and taste threshold data*; ASTM Data Series DS 48A.
15. Baker, R. A., 1963. Threshold odors of organic chemicals. *J Am Water Works*. 55:913–916.

chapter four

Analysis of musty microbial metabolites by stir bar sorptive extraction

Ray Marsili

Contents

The occurrence of objectionable tastes and odors in potable water is widespread. The most troublesome odors that occur usually are those described as muddy, earthy, or musty. Metabolites from microorganisms that contaminate drinking water systems are usually the cause of such tastes and odors. The malodorous metabolites have long been attributed to geosmin (*trans*-1,10-dimethyl-*trans*-9-decalol) and 2-methylisoborneol (2-MIB). Another musty compound that has been reported in the literature is 2-isopropyl-3-methoxy pyrazine (IPMP) [1].

These metabolites are produced by numerous actinobacteria (e.g., *Streptomyces*) and some blue-green algae. Actinobacteria are a group of gram-positive bacteria. They can be terrestrial or aquatic. Human threshold sensitivities for these compounds are extremely low—approximately 10 ng/L for geosmin and 30 ng/L for MIB. The compound IPMP can also be produced by *Harmonia axyridis* (the multicolored Asian Lady Beetle). "Ladybug taint" (LBT) from IPMP contamination has resulted in significant economic losses for vineyards and wineries [2]. Musty off-flavor contamination from these chemicals has been reported for water, beet sugar [3], wine [4], potatoes [5], fish [6], and many other foodstuffs.

Another class of off-flavor compounds that can lead to musty off-flavors in water, and food and beverage products, is trihaloanisoles. Contamination of wine with fungal aromas is considered one of wine industry's worst threats. The most commonly reported culprit is 2,4,6-trichloroanisole (2,4,6-TCA), a compound long associated with the wine industry's cork taint problem.

The extent of losses from trihaloanisole cork taint is difficult to determine accurately, with estimates depending on whether a study originates from a cork-related sector, the wineries, or companies that make synthetic stoppers. According to one estimate, a retail value of $180–$630 million of California wines is spoiled each year by cork taint [7]. The cork-industry group APCOR cites a study showing a 0.7%–1.2% taint rate. However, other studies suggest the taint rate could be much higher. In a 2005 study of 2800 bottles tasted at the *Wine Spectator* [8] blind-tasting facilities in Napa, California, 7% of the bottles were found to be tainted.

As shown in Table 4.1, many types of foods and beverages can become contaminated with trihaloanisoles. For example, casein, a widely used dairy-based ingredient, was recently implicated in significant product recalls with several foods.

Table 4.2 shows the most common bacterial and fungal metabolites that contribute earthy, musty, and muddy off-flavors to water, food products, and beverages.

Table 4.1 Examples of Foods Contaminated by Haloanisoles

Musty contaminant	Type of food	Researcher/year
TCA	Wine	First reported by Tanner 1981 [9]; Buser 1982 [10]
2,3,4,6-TeCA[a]	Chickens and eggs	Engel 1966 [11]
2,3,4,6-TeCA	Chickens	Curtis 1972 [12], 1974 [13]
Dibromoanisole	Casein	Cant 1979 [14]
2,4,6-TCA and 2,3,4,6-TeCA	Dried fruits, packaging materials	Whitfield 1985 [15]; Tindale 1989 [16]
2,4,6-TCA and 2,3,4,6-TeCA	Brazilian coffee	Spadone 1990 [17]
2,4,6-TCA	Drinking water	Nystrom 1992 [18]
2,4,6-TCA	Sake (Japanese wine)	Miki 2005 [19]
2,4,6-TCA and 2,3,6-TCA	Casein, chocolate beverage	Marsili 2007 (unpublished)
2,4,6-TBA[b]	Casein, nondairy creamer	Marsili 2007 (unpublished)
2,4,6-TBA	Casein	Andrewes 2010 [20]

[a] 2,3,4,6-Tetrachloroanisole.
[b] 2,4,6-Tribromoanisole.

Table 4.2 Metabolites That Contribute Musty Taints

Musty compounds	Odor threshold	Odor type	Molecular mass
2-MIB[a] (2-methyl isoborneol)	5–10 ng/L	Earthy	152
Geosmin[b]	1–10 ng/L	Musty	182
2,4,6-TCA[c] (2,4,6-trichloroanisole)	0.1–2 ng/L	Musty	212
2,3,6-TCA[c] (2,3,6-trichloroanisole)	0.1–2.0 ng/L	Musty	212
2,3,4-TCA[c] (2,3,4-trichloroanisole)	0.2–2.0 ng/L	Musty	212
2,4,6-TBA[c] (2,4,6-tribromoanisole)	0.15–2.0 ng/L	Musty	346
IPMP[d] (2-isopropyl-3-methoxy pyrazine)	0.3–2.3 ng/L (in wine)	Musty	152

[a] Cyanobacteria (blue-green algae).
[b] Cyanobacteria and actinobacteria (especially *Streptomyces*).
[c] Haloanisoles: from halophenols by biological action of fungi (*Trichoderma longibrachiatum*) by biomethylation reaction (chlorophenol *O*-methyltransferase).
[d] From actinobacteria (*Streptomyces*) and from multicolored Asian Lady Beetle.

Analytical approaches to monitoring nanogram-per-liter levels of musty metabolites

Efforts to monitor and control geosmin, MIB, IPMP, and trihaloanisoles require sensitive analytical tools and techniques. Some researchers have been successful in applying solid-phase microextraction (SPME) techniques. For example, the application of SPME and gas chromatography–mass spectrometry (GC–MS) with initial cool programmable temperature vaporizer (PTV) inlet has been reported to work well [21]. We found stir bar sorptive extraction (SBSE) in combination with GC–time-of-flight MS (GC–TOFMS) to be an excellent analytical strategy.

Analytical methods to extract and concentrate malodorous contaminants that are often present in trace levels in food products and beverages often include liquid–liquid extraction (LLE) and solid-phase extraction (SPE). These methods suffer from excessive solvent use and/or the inability to obtain significant analyte enrichment, requiring concentration to smaller volumes (e.g., <1 mL) or large volume injection into the GC inlet. To compensate for these inadequacies, solventless methods like SPME and SBSE, which permit extraction and concentration in a single analytical step, were developed and they have become popular in recent years. Both SPME and SBSE quantitatively introduce analytes into the GC column by thermal desorption. However, since SBSE has a 50–250 times larger volume of extraction phase coating compared with SPME, SBSE-enrichment factor is higher than SPME, allowing extremely low limits of detection (LODs) at the sub-nanogram per liter level, particularly for solutes having hydrophobic characteristics.

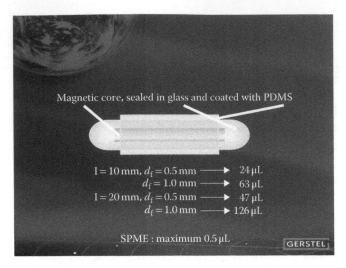

Figure 4.1 The GERSTEL Twister: Automated extraction from liquid samples.

The SBSE recovery can be estimated if the octanol–water distribution coefficient (K_{ow}) of the analyte is known. The coefficient K_{ow} is the ratio of the concentrations of an analyte in octanol and in water at a specified temperature. The musty metabolites listed in Table 4.2 have relatively high K_{ow} values and, therefore, extract with high recoveries into a GERSTEL Twister. Hydrophilic solutes with low values of K_{ow} show lower recovery. However, the sequential SBSE developed by Ochiai et al. [22] can significantly improve extraction recoveries for hydrophilic analytes.

The SBSE technique is based on the GERSTEL Twister, which consists of stir bars coated with polydimethylsiloxane (PDMS). As shown in Figure 4.1, the Twister is offered in four versions. The SBSE technique works on the same principle as LLE. Extraction is based on sorption. The technique allows predictable recoveries based on log K_{ow} of the analytes. The PDMS is an excellent absorbent because it is highly inert, does not retain water, has background chemicals that are readily identifiable, and selectively eliminates polar matrix interferences, which is an advantage when analyzing alcoholic beverages as ethanol is not readily absorbed by PDMS.

Stir bar sorptive extraction of geosmin in drinking water

A municipal water treatment facility located in the Pacific Northwest submitted three samples for analysis. One sample was a control (with no malodor) and two were complaint samples having musty malodors. Two

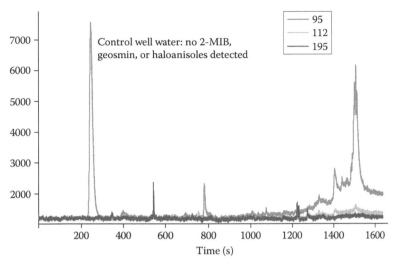

Figure 4.2 Stir bar sorptive extraction (SBSE) with 1 cm × 0.5 mm Twister, 10 mL of control water, 1 h extraction at 1000 rpm stirring; gas chromatography–time-of-flight mass spectrometry (GC–TOFMS).

outside laboratories were unable to detect any of the musty metabolites listed in Table 4.2 in the complaint samples. The samples were analyzed by SBSE with a 1 cm × 0.5 mm Twister (Figure 4.2). A total of 10 mL of water was extracted for 1 h at a stirring rate of 1000 rpm. Due to its high sensitivity and ability to perform peak deconvolution, a Leco Pegasus 4D was selected as the GC–MS system. None of the musty metabolites listed in Table 4.1 were detected in the control water. As shown in Figure 4.3, the same analytical method was used for one of the complaint samples. A peak was detected for geosmin at 1013 s. However, the signal-to-noise ratio (S/N) obtained of 5.0 is rather low.

Additional experiments were conducted to confirm the identity of the geosmin peak in the complaint sample. First, a standard mixture was prepared by adding 500 ng/L of 2-MIB; 2,4,6-TCA; and geosmin to the control water sample. As shown in Figure 4.4, these peaks were easily identified in the chromatogram. The retention time of the geosmin peak in the standard was identical to the peak identified as geosmin in the complaint sample.

Finally, changes were made in the method to improve sensitivity. A larger Twister (2 cm × 0.5 mm) was used in addition to a larger sample weight of water (25 g). The extraction time was increased to 2 h. The geosmin peaks in the complaint samples were clearly discernible. Note the improvement in S/N in Figure 4.5. The concentrations of geosmin in the complaint samples were 21.6 and 21.2 ng/L. The quantitation mass used for geosmin was $m/z = 112$.

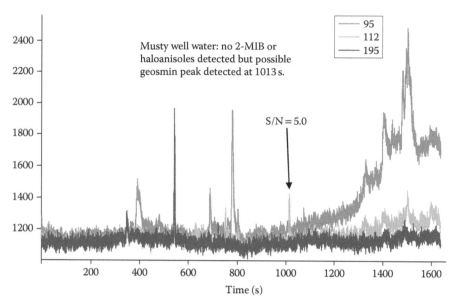

Figure 4.3 SBSE with 1 cm × 0.5 mm Twister, 10 mL of complaint water, 1 h extraction at 1000 rpm stirring; GC–TOFMS.

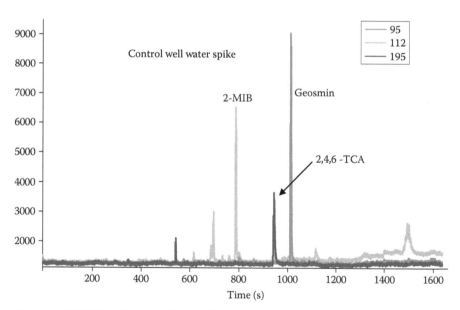

Figure 4.4 SBSE with 1 cm × 0.5 mm Twister, 10 mL of water spiked with 500 ng/L 2-methylisoborneol; 2,4,6-trichloroanisole; and geosmin, with 1 h extraction at 1000 rpm stirring; GC–TOFMS.

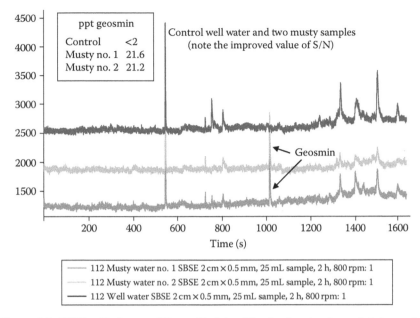

Figure 4.5 SBSE with 2 cm × 0.5 mm Twister, 25 mL of water (complaint samples and control), 2 h extraction at 1000 rpm stirring; GC–TOFMS.

Stir bar sorptive extraction of geosmin in salted snacks

Water is a simple matrix ideally suited to SBSE analysis. A more challenging sample matrix is snack food. A snack food company used to get complaints in the late spring and early summer seasons that their pretzel product had a musty off-flavor. Control and complaint pretzel samples were screened for the musty metabolites listed in Table 4.2. Detection of nanogram-per-liter levels of geosmin in a relatively complicated sample matrix like snack food is considerably more challenging than detection in water.

The SBSE GC–TOFMS analysis of complaint pretzels showed that the cause of the malodor was geosmin. A total of 20 g of pretzel sample was macerated with 100 mL of 15% (v/v) ethanol using a handheld blender for 3 min (Figure 4.6). The sample was centrifuged. A total of 20 mL of centrifugate was added to a flat-bottomed flask and stirred using a 2 cm × 0.5 mm Twister for 2 h at 1000 rpm. The Twister was then thermally desorbed in a GERSTEL thermal desorption unit (TDU) system. Analysis was conducted with selected ion monitoring at $m/z = 112$ with $m/z = 182$ as a confirmation ion. Quantitation was made based on the response factor determined from peak area results of control samples spiked with 0-, 50-, 100-, 300-,

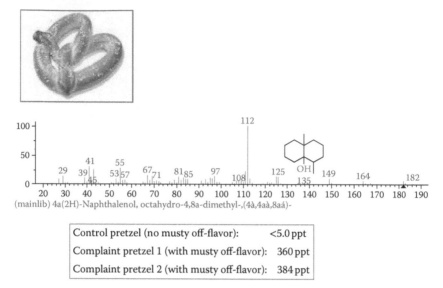

(mainlib) 4a(2H)-Naphthalenol, octahydro-4,8a-dimethyl-,(4à,4aà,8aá)-

Control pretzel (no musty off-flavor):	<5.0 ppt
Complaint pretzel 1 (with musty off-flavor):	360 ppt
Complaint pretzel 2 (with musty off-flavor):	384 ppt

Figure 4.6 Analysis of complaint pretzels: 20 g of pretzel was extracted with 100 mL of 15% ethanol with a handheld blender (3 min). Then 20 mL of centrifugate was collected and extracted with 2 cm × 0.5 mm Twister for 2 h at 1000 rpm. Mass spectrum of geosmin peak in the complaint pretzel is shown.

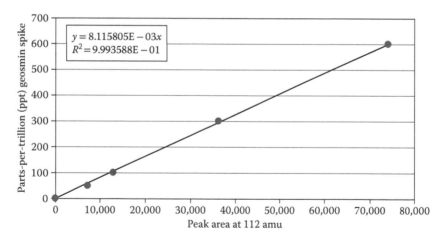

$$y = 8.115805E - 03x$$
$$R^2 = 9.993588E - 01$$

Figure 4.7 Geosmin standard calibration curve (spike in control pretzel).

and 600-ppt geosmin (Figure 4.7). The linear least-squares correlation coefficient of the calibration curve was 0.9994. Note that each calibration point uses a different Twister for each sample.

Complaint pretzels were contaminated with geosmin in the range of 350–400 ppt. Testing of processing water revealed that the source of the geosmin in the samples was the processing water from the local municipal water treatment facility. During periods of heavy rainfall, algal blooms occurred in the water and algal metabolites were the source of geosmin. A charcoal filtration facility for the filtration of water was installed to eliminate the problem.

Novel approach to collecting musty off-flavor compounds on tap

Researchers at Veolia Environmental (Paris, France), one of the largest water suppliers in the world, recently developed Advanced Relevant Investigation Sampler for Taste and Odor at Tap (ARISTOT), which allows true passive sampling of off-flavors directly from the consumers' water faucet [23]. The patented passive sampler, which consists of a metal cartridge that is loaded with seven PDMS-coated stir bars, screws onto the faucet and extracts and concentrates off-flavor compounds directly from the consumers' water tap under standard usage conditions without the use of extracting solvents. The extracted compounds absorbed onto the Twisters are subsequently mailed back to the analytical laboratory for determination of ultratrace levels of malodorous compounds using thermal desorption GC–MS. With this technique, it is possible to detect geosmin, 2-MIB, and haloanisoles in the low parts-per-quadrillion (pictogram per liter) concentration range.

Haloanisole contamination of casein

Casein powder ingredients are commonly used in food and beverage applications because of their flavor stability and functional properties. Casein powders are used in cheese analogues, bakery products, meat products, confectionery products, desserts, nondairy coffee creamers, and beverages. Casein has a characteristic unpleasant stale flavor, often described as "animal/wet dog," "stale," "gluelike," or "burnt feather" [24]. A similar flavor has been found in other dairy products such as stored skim milk powder [25] and sterilized concentrated milk [26]. o-Aminoacetophenone has been identified as an important aroma component of stored skim milk powders [27]. More recently, however, researchers using sensory analysis of model aroma systems determined that the typical malodor of rennet casein was principally caused by hexanoic acid, indole, guaiacol, and p-cresol [28].

In the summer of 2007, our laboratory received several complaint casein powder samples from three different international food/beverage companies. All casein samples suffered from an extreme musty taint and had been implicated as the cause of objectionable malodors in processed food samples—including cereal bars, nutritional beverages, and a chocolate-flavored beverage—returned to the manufacturers as consumer complaints. Studies were conducted to determine if the malodor was simply the typical casein malodor, albeit in an abnormally high level, or if the musty malodor originated from one or more of the microbial metabolites listed in Table 4.1. Both tainted products and the caseins associated with the products were analyzed.

Since numerous types of malodor chemicals were to be measured at low detection limits, sequential SBSE GC–TOFMS was investigated as a possible technique. First, extraction from aqueous solution was investigated. In this method, 1 g of the complaint casein sample was vigorously stirred with 25 mL of distilled water for 30 min at room temperature with a Teflon-coated micro stir bar. After 30 min, the Teflon-coated stir bar was removed and replaced with a 2 cm × 0.5 mm Twister (PDMS coated). The sample was then stirred at 1000 rpm for 3 h prior to thermal desorption into a GC–TOFMS. A second 1-g sample of the complaint casein was then extracted using the same method, only 2 g of sodium chloride was added prior to adding the Twister stir bar. This would measure which chemicals experienced significant extraction recoveries by salting out. A third 1-g sample of the complaint casein was then extracted after low pH adjustment by adding 5 mL of sodium dihydrogen phosphate (25% w/v) buffer. This experiment determined if there were analytes (e.g., volatile organic acids) better recovered by SBSE at lower pH levels.

The results of these experiments showed that some flavor-important analytes extract better into Twister in aqueous solution (e.g., trihaloanisoles; Figure 4.8), whereas other analytes (e.g., 4-amino acetophenone, pyrazines, and phenylethyl alcohol; Figure 4.9) were extracted with higher recoveries after salting out.

The SBSE testing of all three complaint casein samples showed that two of the samples were contaminated with 2,4-dichloroanisole; 2,4,6-TCA; and 2,3,6-TCA, and one sample was contaminated with 2,4,6-tribromoanisole (2,4,6-TBA). The mechanism for the formation of 2,4,6-TCA from 2,4,6-trichlorophenol is shown in Figure 4.10. Trichlorophenol is a common ingredient of fungicides, pesticides, and wood preservatives. The 2,4,6-TBA detected in the third complaint casein sample likely originated from a similar mechanism, only 2,4,6-tribromophenol (2,4,6-TBP), used in flame retardants and fungicides, was the likely substrate.

A chromatogram of a complaint casein sample with TCA contamination is shown in Figures 4.11 and 4.12. Using Leco's ChromaTOF software, the chromatogram in Figure 4.11 can be replotted at masses 212, 195, 161, and 196. As shown in Figure 4.12, the mass spectra of the resulting peaks can be compared with the National Institute of Standards and Technology (NIST) library mass spectra to quickly identify the chloroanisoles.

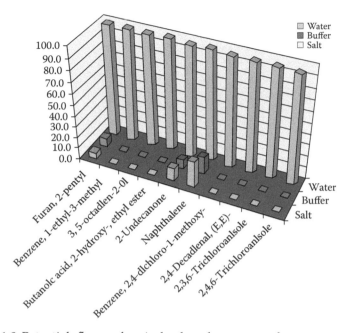

Figure 4.8 Potential flavor chemicals that demonstrate best recovery when extracted by Twister in aqueous solution.

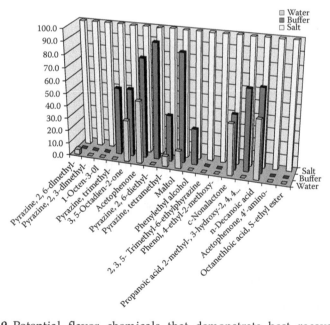

Figure 4.9 Potential flavor chemicals that demonstrate best recovery when extracted by Twister in saturated salt solution.

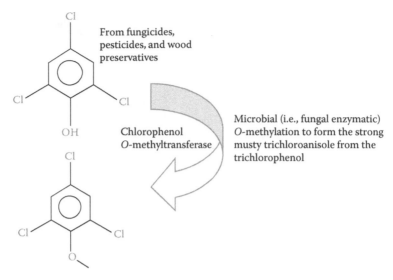

Figure 4.10 Mechanism for the formation of 2,4,6-trichloroanisole from 2,4,6-trichlorophenol.

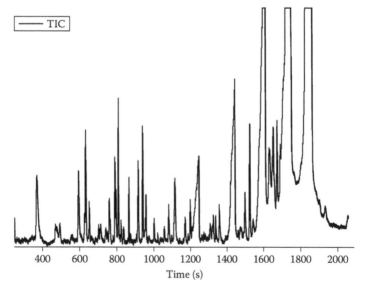

Figure 4.11 Chromatogram of a complaint casein sample with trichloroanisole contamination: 1 g of musty casein + 25 mL of water is taken; stir for 30 min at room temperature. SBSE for 3 h with 2 cm × 0.5 mm Twister is performed. TIC = total ion chromatogram.

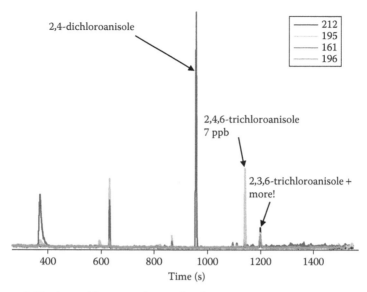

Figure 4.12 Finding chloroanisoles in the chromatogram.

One significant advantage of using Leco's GC–TOFMS instrument is its ability to perform peak deconvolution. Although peak deconvolution software exits for mass selective detection (MSD) data, TOFMS, a continuous array analyzer compared with the sequential scanning MSD, has three important advantages over the quadrupole MS detector: (1) much higher data density since spectra are collected at an acquisition rate of up to 500 scans·s compared to the 5–10 spectra·s rate for quadrupoles; (2) no spectral skewing even for fast eluting peaks; and (3) ease of performing peak deconvolution. These advantages have been discussed in detail in the literature [29].

Since a single sharp chromatographic peak occurs at a retention time of 1200 s, the peak appears to be 2,3,6-TCA. However, as shown in Figure 4.13, peak deconvolution actually results in three different chemicals that are coeluting at nearly the same retention time: (1) 2,3,6-TCA; (2) 2,4,6-trichlorophenol; and (3) butyl butanoate. This result was extremely significant because it showed that the substrate 2,4,6-trichlorophenol from which the 2,4,6-TCA originated was present in the casein sample. This finding strongly supported the proposed mechanism of formation shown in Figure 4.10. This information was important in law suit resolution.

Modifications of the TCA procedure were investigated to improve detection limit. Prestirring the casein–water mixture at elevated temperature (50°C instead of room temperature [RT]) and using 3 g of casein in

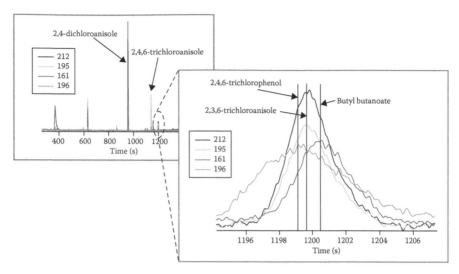

Figure 4.13 Value of peak deconvolution.

35 mL of water was investigated. As shown in Figure 4.14, no significant increases in TCA peak areas occurred. Therefore, this test modification was not incorporated into the method.

Adding organic modifiers was shown to be an effective way of increasing analyte recovery for some analytes analyzed by SBSE. For example, Serodio and Nogueira [30] showed that addition of 5% methanol modifier to water samples prior to SBSE significantly improved the recoveries of eight pyrethroid pesticides in water samples. Figure 4.15 shows that attempts to increase the recovery of TCA peaks with the addition of ethanol or ethyl acetate as organic modifiers did not improve test detection limits. Therefore, the original method was used without modification.

The SBSE GC–TOFMS testing of the third complaint sample revealed 2,4,6-TBA contamination. This casein sample was sourced from India. The SBSE GC–TOFMS procedure used to detect TCA in casein was employed in this case also. However, in this case, quantitation was based on spiking the samples with isotopically labeled 2,4,6-TBAd (2,4,6-tribromoanisole, deuterated).

Analyte recoveries at low levels from challenging matrices can be significantly less than 100% and be quite variable. One technique to compensate for these variations is to add a surrogate standard to each sample analyzed. The chemical selected as a surrogate standard must behave in the same way as the analyte during sample preparation. If the analyte is ionizable, for example, the surrogate should have an identical pK_a. It should have identical polarity as the analyte, but it should not occur naturally in the samples. Selecting a proper surrogate based on these criteria is

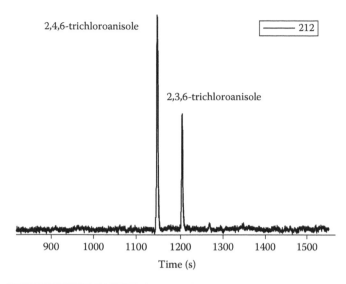

Sample	2,4,6-	2,3,6-
1 g/25 mL water, stir at RT for 30 min	348,765	140,852
1 g/25 mL water, stir at 50°C for 30 min	325,520	158,456
3 g/35 mL water, stir at RT for 30 min	312,502	152,591
3 g/35 mL water, stir at 50°C for 30 min	216,307	112,271

Figure 4.14 SBSE method development: The effect of prestirring at room temperature and at 50°C and the effect of using larger sample weights of casein are shown. Numbers in the table denote peak areas ($m/z = 212$) of the two different trichloroanisole isomers. Samples were extracted at 3 h with 2 cm Twister in all cases.

difficult with GC detectors other than a mass spectrometric detector. With mass spectrometric detection, it is possible to use chemical surrogates that satisfy these criteria. They are analogues of the analyte itself differing only in their mass. These analogues of the analytes of interest are prepared by replacing one or more atoms in the molecule with heavier, stabler isotopes—usually replacing ^1H with ^2H atoms or ^{12}C with ^{13}C atoms. In all other aspects, analytes and their stable isotope analogues are identical. Prior to extraction, quantifiable amounts of these stable isotope surrogates are added to the sample. A poor recovery of the analyte from a particular sample matrix will be exactly matched by a correspondingly poor recovery to the surrogate standard, enabling corrections and accurate analytical results of the analyte. The downside of using stable isotope surrogates is the cost associated with their purchase or synthesis.

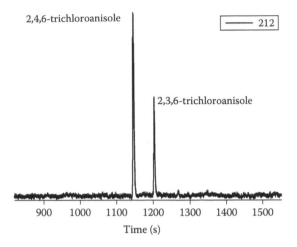

Sample	2,4,6-	2,3,6-
1 g/25 mL water, stir at RT for 30 min	184,787	97,712
1 g/25 mL 10% EtOH, stir at 50°C for 30 min	130,826	64,587
1 g/25 mL 5% EtOAc, stir at RT for 30 min	None detected	None detected

Figure 4.15 SBSE method development: The effect of extracting samples with added organic solvents is shown. Numbers in the table denote peak areas ($m/z = 212$) of the two different trichloroanisole isomers. Samples extracted at 3 h with 2 cm Twister in all cases.

A chromatogram of a complaint sample contaminated with 2,4,6-TBA spiked with deuterated 2,4,6-TBA at 133 ppb is shown in Figure 4.16. This quantitation technique is sometimes referred to as stable isotope dilution analysis (SIDA). It is arguably the most accurate technique for quantitative determination of organic compounds by GC–MS analysis [31, 32]. The level of 2,4,6-TBA measured in the complaint samples was approximately 200 ppb. The standard calibration curve shown in Figure 4.17 was generated by spiking control casein having no detectable TBA with various levels of TBAd standard.

The company that supplied the TBA-tainted casein ingredient to various food processors later published an article confirming our results. This article [33] explained a possible mechanism for contamination of the casein ingredient.

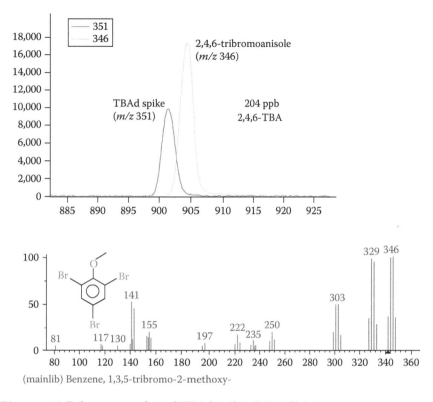

(mainlib) Benzene, 1,3,5-tribromo-2-methoxy-

Figure 4.16 Tribromoanisole and TBAd spikes (133 ppb) in musty casein.

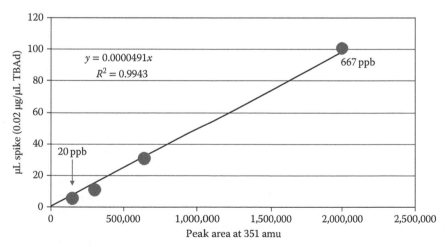

Figure 4.17 TBAd standard curve in control casein powder (3 g).

Only a few cases of TBP and TBA tainting have been reported in the literature, and this is the first report of tainting by these compounds in a dairy product. This work reported that TBA concentration in a tainted beverage made with TBA-tainted casein contained between 20 and 560 ppb of TBA, amounts far exceeding the sensory threshold level for 2,4,6-TBA.

A previous instance of caseinate tainting with a bromoanisole compound was reported for α,α-dibromoanisole that occurred during manufacture [34]. This compound, which contributed an unusual iodine-like off-flavor, was likely formed from the reaction at high pH between phenol and bromoform (produced by the reaction at high pH between acetone, bromide ion, and hypochlorite sanitizer). However, with the TBA-tainted casein, the offending chemical likely resulted from an initial contamination with the fungicide TBP followed by microbial enzyme conversion to TBA. In all cases, TBA was the only musty compound detected in complaint samples of calcium caseinate, which was found in association with its precursor TBP. On reproducing the musty defect by spiking control calcium caseinate with TBA at the concentration levels measured in the tainted product, TBA was proved to be the causative agent of the musty defect in the calcium caseinate.

The TBA-treated lumber used to construct wood floor in a shipping container was initially suspected as the source of the TBP precursor. However, testing of numerous timber materials throughout the company's global supply chain failed to detect significant levels of TBP. The search for the source of contamination was expanded to include the slip sheets that are commonly used to slide products into shipping containers instead of loading on wooden pallets. These ribbed plastic sheets were manufactured from recycled plastics. Calcium caseinate is typically stacked on to a slip sheet that is 2.24×1.38 m, 1-mm thick, and contains approximately 2 kg of high-density polyethylene plastic resins.

Because food products are not in direct contact with the slip sheets, the use of carefully selected recycled plastics for slip sheet construction was considered unnecessary. Thus, TBP was detected in most of the slip sheet samples tested. The slip sheets did not smell musty, nor was any TBA detected in them. For the slip sheets to be a source of both TBP and TBA, some of the TBP (that migrates from the sheets) must be converted to TBA by microorganisms in the general environment. The source of TBP in the recycled plastics is unknown, but it is likely associated with the use of organobromine flame retardants such as polybrominated diphenyl ethers (PBDEs) in plastics. It is surprising that TBP could leach from the plastic slip sheets, pass through paper bags, and contaminate the casein powder. This example of a food taint illustrates the extraordinary vigilance that is required to prevent off-flavor issues in foods and beverages.

Headspace sorptive extraction gas chromatography–time-of-flight mass spectrometry multivariate analysis for chemotaxonomic profiling of mold metabolites

Slight variations of the analytical techniques reported in this chapter for musty mold metabolites have been applied to the study of other microbial metabolite problems. For example, chemotaxonomic classification of molds based on their metabolite profiles obtained by headspace sorptive extraction (HSSE) GC–TOFMS profiles has been reported [35]. In this study, volatiles from fungi cultivated in petri dishes were collected by a simple headspace PDMS sorptive extraction technique, thermally desorbed into a GC capillary column, and identified by GC–TOFMS. Metabolite profiles of seven different species of fungi grown on two types of sterile agars (potato dextrose and Sabouraud dextrose) were determined and subjected to multivariate analysis (principal component analysis [PCA]) to determine whether separate classes of fungi can be distinguished from one another based on their metabolite profiles. Three species from the genus *Penicillium* and four out-groups each from a different phylum were studied. The species studied are listed in Table 4.3.

An earlier analytical approach for collecting volatile fungi metabolites was developed by Larsen and Frisvad [36]. The method was based on diffusive sampling from petri dishes. Adsorption tubes filled with either carbon black or Tenax TA were placed in the headspace of a petri dish that had been incubated with mold. The volatiles diffused into the tubes and were trapped by the adsorbent. Organic volatiles were then eluted off the traps with 1.5 mL of diethyl ether and injected into a GC–MS system. A drawback of this diffusive sampling method is that tube openings are small compared with the surface area of fungi in the petri dishes. The diffusive sampling method and the HSSE method are compared in Figure 4.18.

With the HSSE approach, more trapping medium is readily exposed to volatiles in the headspace of the petri dish, greatly facilitating volatile absorption. Another advantage of the HSSE technique is that it is a solventless technique, it reduces the possibility of contamination, and it avoids the nitrogen blowdown step of the diffusive sampling method, which is time consuming and risks potential loss of volatile metabolites that could occur during solvent evaporation.

After 5 days of inoculation and incubation at 25°C sufficient mold growth was observed on all plates to conduct HSSE of mold metabolites, with the exception of a few mold cultures that failed to grow on the potato dextrose agar. These were *Coprinus cinereus* and *Rhizopus stolonifer*. All species grew successfully on the Sabouraud dextrose agar, making it a more

Table 4.3 Taxonomic Classification of *Penicillium* Species and Out-Groups Based on GC–TOFMS Metabolite Profiles

Kingdom	Phylum	Class	Order	Family	Genus	Species	Type
Fungi	Deuteromycota	Eurotiomycetes	Moniliales	Moniliaceae	*Penicillium*	*roqueforti*	Cheese mold
Fungi	Deuteromycota	Eurotiomycetes	Moniliales	Moniliaceae	*Penicillium*	*camemberti*	Cheese mold
Fungi	Deuteromycota	Eurotiomycetes	Moniliales	Moniliaceae	*Penicillium*	*italicum*	Plant mold
Fungi	Zygomycota	Zygomycetes	Mucorales	Mucoraceae	*Rhizopus*	*stolonifer*	Filamentous mold
Fungi	Basidiomycota	Homobasidiomycetes	Agaricales	Agaricaceae	*Coprinus*	*cinereus*	Mushroom
Fungi	Ascomycota	Ascomycetes	Sphaeriales	Sordariaceae	*Sordaria*	*fimicola*	Microfungus
Protocista	Heterkonta	Oomycotea	Saprolegniales	Saprolegniaceae	*Saprolegnia*	sp.	Water mold

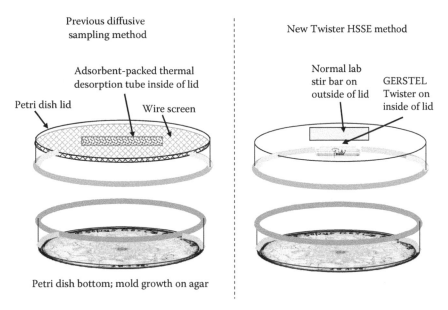

Figure 4.18 Extraction of fungal metabolites by two different methods: Diffusive sampling with adsorbent tubes and headspace sorptive extraction (denoted as HSSE in the figure).

suitable agar for a large array of fungi. The GERSTEL Twister stir bar was placed 1.0 cm from the surface of the agar medium and allowed to absorb volatile metabolites for 90 min. The volatiles trapped on the Twister were desorbed with a GERSTEL TDU using splitless desorption. Desorbed volatiles were cryofocused with liquid nitrogen in the GERSTEL CIS4 inlet at −60°C, with a carrier gas flow of 50 mL/min. After cryofocusing was completed, the volatiles were transferred into the capillary column by heating the CIS4 inlet at a rate of 10°C/s to 300°C (3-min hold) keeping the inlet splitless for 1.5 min.

Chromatographic peak identification was made by library matching using the 2005 NIST MS Library and Leco's ChromaTOF software. Some of the fungal metabolites measured appear in Table 4.4. Each sample contained over 1000 volatiles, but only the largest peaks are shown in Table 4.4. Peak areas for 118 volatile metabolites were selected for PCA analysis.

Figure 4.19 shows a chromatogram for *C. cinereus* grown on Sabouraud dextrose. Of all the species studied *C. cinereus* produced the most volatile metabolites, many of which were sesquiterpenes that were not detected at significant levels in any of the samples. Two of the out-group species, *R. stolonifer* and *Sordaria fimicola*, had the highest levels of phenyethyl alcohol, a compound having a floral/rose odor. *Saprolegnia* sp. had the highest

Table 4.4 Examples of Fungal Metabolites Detected by HSSE GC–TOFMS, Where x Is a Small Analytical Peak, xx Is a Medium-Size Peak, and xxx Is a Large Analytical Peak

Analyte	CAS number	Out-groups							*Penicillium* species			
		#1 S	#2 S	#3 S	#4 S	#5 P	#6 P	#7 S	#8 P	#9 S	#10 P	#11 S
2-Heptenal, (E)-	18829-55-5	0	0	0	0	0	x	x	x	x	x	x
Decane	124-18-5	xxx	xxx	xxx	xxx	xxx	0	xxx	0	0	0	xxx
Phenol	108-95-2	0	0	0	0	0	xx	0	0	x	x	0
Pentanoic acid	109-52-4	0	0	xxx	0	x	0	xxx	0	0	0	0
Nonane, 2,6-dimethyl-	17302-28-2	xxx	0	xxx	0	0	0	0	0	0	0	0
trans-3-caren-2-ol	0-00-0	0	0	0	xx	0	0	0	0	0	0	0
3-Pentanol	584-02-1	xxx	0	0	0	0	0	xx	0	0	0	xx
Decane, 4-methyl-	2847-72-5	xxx	xxx	xxx	xxx	xxx	0	0	0	0	0	0
Thujone	546-80-5	0	x	0	0	0	0	0	0	0	0	0
cis-Geraniol	106-25-2	0	0	0	x	0	0	0	0	0	0	0
Undecane	1120-21-4	xxx	xxx	xxx	xxx	xxx	0	x	0	x	0	x
Linalool	78-70-6	xxx	0	0	0	0	0	x	0	x	xx	xx
Phenylethyl alcohol	60-12-8	0	0	xxx	xxx	xxx	x	x	0	0	xx	xx
Megastigma-4,6(E),8(Z)-triene	71186-24-8	xxx	x	0	0	0	0	0	0	0	0	0
α-Gurjunene	489-40-7	xxx	0	0	0	0	0	0	0	0	0	0
Neoclovene	4545-68-0	xx	0	0	0	0	0	0	0	0	0	0
Thujopsene-(I2)	0-00-0	0	xx	0	0	0	0	0	0	0	0	0
Indole	120-72-9	0	0	0	0	0	x	0	0	0	x	0
Chamigrene	18431-82-8	xxx	0	0	0	0	0	0	0	0	0	0
(+)-Sativene	3650-28-0	xxx	xxx	0	0	0	0	0	0	0	0	0
Di-epi-alpha-cedrene-	21996-77-0	xxx	x	0	0	0	0	0	0	0	0	0

Compound	CAS	#1	#2	#3	#4	#5	#6	#7	#8	#9	#10	#11
Viridiflorol	552-02-3	xxx	o	o	o	o	o	o	o	o	o	o
Santene	529-16-8	xxx	o	o	o	o	o	o	o	o	o	o
Allo-aromadendrene	25246-27-9	xxx	o	o	o	o	o	o	o	o	o	o
β-Caryophyllene	87-44-5	xxx	o	o	o	o	o	o	o	xxx	o	o
Cedrene	11028-42-5	x	o	o	o	o	o	o	o	o	o	o
Elemene	339154-91-5	x	o	o	o	o	o	o	x	o	o	o
α-Caryophyllene	6753-98-6	xx	x	o	o	o	x	x	o	o	o	o
α-Bergamotene	17699-05-7	o	o	o	o	x	x	x	x	o	o	x
Caryophyleine-(13)	136296-37-2	o	o	o	o	o	o	o	x	xxx	xxx	o
cis-a-Bisabolene	29837-07-8	xxx	x	o	o	o	o	o	o	xxx	xxx	o
Aristolochene	0-00-0	xxx	x	o	o	o	o	o	o	x	o	x

Note: S = Sabouraud dextrose; P = potato dextrose; #1 = C. cinereus; #2 = Saprolegnia sp.; #3 = R. stolonifer; #4 and #5 = S. fimicola; #6 and #7 = P. italicum; #8 and #9 = P. camemberti; #10 and #11 = P. roqueforti. Where #1 = Coprinus cinereus (S-Dex); #2 = Saprolegnia Sp. (S-Dex); #3 = Rhizopus stolonifer (S-Dex); #4 = Sordaria fimricola (P-Dex); #5 = Sordoria fimricola (S-Dex); #6 = Penicillium italicum (P-Dex); #7 = Penicillium italicum (S-Dex); #8 = Penicillium camemberti (P-Dex); #9 = Penicillium camemberti (S-Dex); #10 = Penicillium roqueforti (P-Dex); #11 = Penicillium roqueforti (S-Dex). — = no metabolite peak (<500,000); xx = medium metabolite peak (500,000 to 1,000,000 cts); xxx = large metabolite peak (>1,000,000 cts).

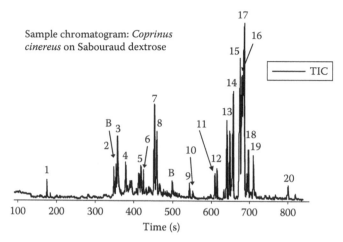

Figure 4.19 Sample chromatogram of volatile metabolites by HSSE GC–TOFMS. (1) acetic acid (agar background); (2) 3-octanone; (3) decane; (4) 2,6-dimethyl nonane; (5) 2-methyl decane; (6) 3-methyl decane; (7) undecane; (8) linalool; (9) octanoic acid, ethyl ester (agar background); (10) decanal (agar background); (11) megastigma-4,6(E),8(Z)-triene; (12) a-gurjunene; (13) chamigrene; (14) longifolene-(V4); (15) aristolochene; (16) sativene + di-epi-a-cedrene; (17) 4-ethyl-4-methyl-2-cyclohexen-1-one; (18) caryophyllene; (19) allo-aromadendrene; (20) *cis*-a-bisabolene; and (B) polysiloxane background peak from Twister. TIC = total ion chromatogram.

level of 1-octen-3-ol (strong mushroom odor). This compound, however, was also detected as a smaller peak in all the *Penicillium* sample chromatograms. The species *P. roqueforti* had the highest level of linalool, a compound with a strong floral odor, of all the *Pencillium* molds. Only *P. camemberti* contained 2-methyl-2-undecanethiol, one of the few sulfur compounds detected in the samples. Production of alkanes was the most significant differentiating factor between the *Pencillium* species and the out-groups.

It is a daunting task to decipher which, if any, of the myriad chemicals generated in GC–MS chromatograms are useful for chemotaxonomic purposes. Data collection in science is often based on many measurements performed on numerous samples. Such multivariate data is commonly analyzed using one or two variables at a time. This approach often fails to discover subtle but significant relationships among all samples and variables. Applying multivariate analysis (also known as chemometrics) can reveal hidden patterns in complex data sets by reducing the information to a more comprehensible format. For example, PCA and hierarchical cluster analysis (HCA) help reduce large data sets into a series of optimized and

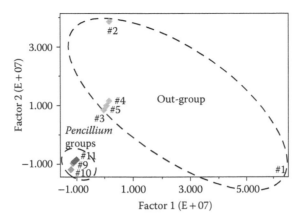

Figure 4.20 PCA plot showing class clustering of *Pencillium* groups and out-groups: Two-dimensional PCA scores plot for peak area data (118 analytes) after meancentering preprocessing was applied.

interpretable views. These views emphasize the natural groupings in the data and show which variables (e.g., fungal metabolite peak areas) most strongly influence the patterns (e.g., classification of samples according to appropriate taxonomic categories). The multivariate analysis software used for this study was Pirouette (Infometrix, Bothell, Washington).

Figure 4.20 is a two-dimensional PCA plot showing that the 118 metabolites compared in all the samples can separate *Penicillium* molds from out-groups in two distinct clusters. Furthermore, an examination of PCA loadings values for factors 1, 2, and 3 shows which metabolites are most influential to clustering. These are listed in Table 4.5. Figure 4.21 shows a magnified view of the *Penicillium* group cluster. This plot shows that the *Penicillium* group cluster actually contains two separate clusters, with the *Penicillium* molds cultured in potato dextrose (#6, #8, and #10) falling in one cluster and the *Penicillium* molds cultured in Sabouraud dextrose falling in the second cluster. The results of PCA show that production of volatiles was highly dependent on species type and slightly dependent on agar type.

Figure 4.22 shows the advantage of peak deconvolution with GC–TOFMS (Leco Pegasus 4D) for detecting peaks that could easily be overlooked with scanning GC–MS instruments. Prior to peak deconvolution, it was difficult to identify peaks in the region of the chromatogram from 691 to 707 s in this particular sample. After peak deconvolution, six metabolite peaks are clearly discernible and accurately integrated.

Table 4.5 PCA Loadings Values for Factors 1, 2, and 3 for Chemicals
Most Influential to Class Clustering

Chemical	CAS number	Factor 1	Factor 2	Factor 3
Decane	124-18-5	0.2790	0.2370	−0.2963
4-Methyl decane	2847-72-5	0.0718	−0.0133	−0.0219
Undecane	1120-21-4	0.2972	0.6110	−0.5162
Linalool	78-70-6	0.1151	−0.0462	0.0378
Phenylethyl alcohol	60-12-8	−0.0002	0.0445	−0.1611
1-Propanone, 1-(5-methyl-2-furanyl)-	10599-69-6	0.0009	0.0385	0.0418
2-Methyl undecane	7045-71-8	0.0133	0.0367	−0.0347
(+)-Sativene	3650-28-0	0.1398	−0.0240	0.0634
2-Cyclohexen-1-one, 4-ethyl-4-methyl-	17429-32-2	0.8560	−0.2926	0.2323
1H-Indene, 1-ethyloctahydro-7a-methyl-	56324-71-1	0.0155	0.6317	0.6859
Methyl cinnamate	103-26-4	0.0269	0.1901	0.2206
Caryophyleine-(I3)	136296-37-2	−0.0535	−0.0954	0.1181
Dodecanoic acid	143-07-7	−0.0478	−0.0563	−0.0037
Phenol, 3-pentadecyl-	501-24-6	0.0022	0.0795	0.0655

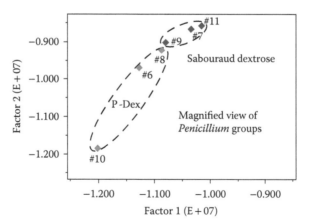

Figure 4.21 Principal component analysis plot showing magnified view of class clustering of *Pencillium* group showing clustering by medium type: Two-dimensional PCA scores plot for peak area data after meancentering preprocessing was applied.

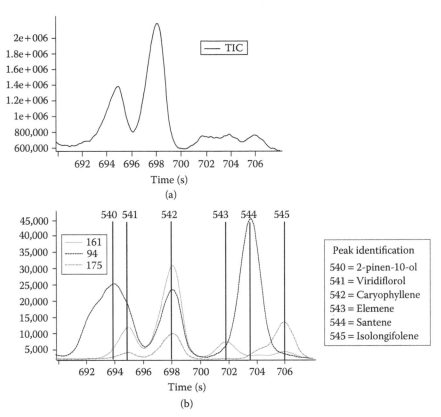

Figure 4.22 The peak deconvolution advantage of GC–TOFMS (a) Total ion chromatogram of *Coprinus cinereus* (Sabouraud dextrose) from 691 to 707 s. (b) The same region of the *C. cinereus* chromatogram after peak deconvolution.

This chapter shows that HSSE GC–TOFMS can be used for generating detailed volatile metabolite profiles that can be used in chemotaxonomic studies of cultured fungi. Furthermore, the application of multivariate analysis and peak deconvolution capabilities of the Leco GC–TOFMS are shown to be powerful problem-solving analytical tools.

References

1. Gerber, N. N. 1977. Three highly odorous metabolites from an actinomycete: 2-isopropyl-3-methoxypyrazine, methylisoborneol, and geosmin. *J Chem Ecol* 3(4):475–482.
2. Pickering, G. J. and Y. Lin. 2006. Asian Lady Beetle (Harmonia axyridis) and wine quality. In *Crops: Growth, quality and biotechnology. III. Quality management of food crops for processing technology,* ed. R. Drish, Helsinki, 785–794. Finland: WFL Publisher.

3. Marsili, R. T., N. Miller, G. J. Kilmer, and R. E. Simmons. 1994. Identification and quantitation of the primary chemicals responsible for the characteristic malodor of beet sugar by purge-and-trap GC–MS-OD techniques. *J Chromatogr Sci* 32(5):165.
4. Pickering, G. J., M. Spink, Y. Kotseridis, I. D. Brindle, M. Sears and D. Inglis. 2008. The influence of *Harmonia axyridis* morbidity on 2-isopropyl-2-methoxypyraine in Cabernet Sauvignon wine. *Vitis* 47(4):227–230.
5. Buttery, R. G., and L. C. Ling. 1973. Earthy aroma of potatoes. *J Agric Food Chem* 21:745.
6. Dupay, H. P., G. J. Flick, A. J. St. Angelo, and G. Sumrell. 1986. Analysis for trace amounts of geosmin in water and fish. *JAOCS* 63(7):905–908.
7. Coque, J. J. R., L. Alvarez Rodriguez, M. Goswami, and R. F. Martinez. 2006. *Causes and origins of wine contamination by haloanisoles (chloroanisoles and bromoanisoles)*. Leon, Spain: Insituto de Biotecnologia. Institute of Biotechnology of Leon (Inbiotec).
8. Laube, J. 2006. Changing with the times. *Wine Spectator,* March 2006, p. 1.
9. Tanner, H., C. Zanier, and C. Wurdig. 1981. Zur analytischen differenzierung von muffon und korkgeschmack in wein. *Schweiz Z Obst U Weinbau* 117:752–757.
10. Buser, H-R., C. Zanier, and H. Tanner. 1982. Identification of 2,4,6-trichloroanisole as a potent compound causing cork taint in wine. *J Agr Food Chem* 30:359–363.
11. Engel, C., A. P. de Groot, and C. Weurman. 1966. Tetrachloroanisole, a source of musty taste in eggs and broiler. *Science* 154:220–271.
12. Curtis, R. F., D. G. Land, M. N. Griffiths, and M. G. Robinson. 1972. 2,3,4,6-Tetrachloroanisole: Association with musty taint in chickens and microbiological formation. *Nature* 235:223–224.
13. Curtis, R. F., C. Dennis, J. M. Gee, M. G. Gee, and M. N. Griffiths. 1974. Chloroanisoles as a cause of musty taint in chickens and their microbiological formation from chlorophenols in broiler house liter. *J Sci Food Agr* 25:811–828.
14. Cant, P. A. E. 1979. Identification of dibromoanisole as a trace flavour contaminant of abnormal commercial casein. *New Zealand J Dairy Sci Tech* 14:35–41.
15. Whitfield, F., L. Nguyen, K. J. Shaw, J. H. Last, C. R. Tindale and G. Stanely. 1985. Contamination of dried fruits by 2,4,6-trichloroanisole and 2,3,4,6-tetrachloroanisole absorbed from packaging materials. *J Agric Food Chem* 45:889–893.
16. Tindale, C. R., F. B. Whitfield, S. D. Levingston, and L. Nguyen. 1989. Fungi isolated fro packaging materials: their role in the production of 2,4,6-trichloroanisole. *J Sci Food Agr* 49:437–447.
17. Spadone, J-C., G. Takeoka, and R. Liadron. 1990. Analytical investigations of Rio off-flavor in green coffee. *J Agr Food Chem* 38:226–232.
18. Nystrom, A., A. Grimvall, C. Krantzrulcker, R. Savenhed, and K. Akerstrand. 1992. Drinking-water off-flavor caused by 2,4,6-trichloroanisole. *Water Sci Technol* 25:241–249.
19. Miki, A., A. Isogai, H. Utsunomiya, and H. Iwata, 2005. Identification of 2,4,6-trichloroanisole (TCA) causing a musty/muddy off-flavor in sake and its production in rice Koji and Moromi mash. *J Biosci Bioengin* 100:178–183.
20. Andrewes, P., J. G. Bendall, G. Davey, and R. Shingleton. 2010. A musty flavour defect in calcium caseinate due to chemical tainting by 2,4,6-tribromophenol and 2,4,6-tribromoanisole. *Int'l Dairy J* 20:423–428.

21. Zhang, L., H. Ruikang and Z. Yang. 2005. Simultaneous picogram determination of "earthy-musty" odorous compounds in water using solid-phase microextraction and gas chromatography–mass spectrometry coupled with initial cool programmable temperature vaporizer inlet. *J Chrom* 1098:7–13.
22. Ochiai, N., D. Sasamoto, H. Kanda, and E. Pfannkoch. 2008. Sequential stir bar sorptive extraction for uniform enrichment of trace amounts of organic pollutants in water samples. *J Chrom A* 1200:72–79.
23. Tondelier, C., T. Thouvenot, A. Genin, and D. Benanou. 2009. On-tap passive enrichment, a new way to investigate off-flavor episodes in drinking water. *J Chrom A* 12(14):2854–2859.
24. Ramshaw, E. H., and E. A. Dunstone. 1969. Volatile compounds associated with the off-flavor in stored casein. *J Dairy Res* 36:215–223.
25. Karagul-Yuceer, Y., K. R. Cadwallader, and M. A. Drake. 2002. Volatile flavor components of stored nonfat dry milk. *J Agric Food Chem* 50:305–312.
26. Arnold, R. G., L. M. Libbey, and E. A. Day. 1966. Identification of components in the stale flavor fraction of sterilized concentrated milk. *J Food Sci* 3:566–573.
27. Parks, O. W., D. P. Schwartz, M. Keeney. 1964. Identification of *o*-aminoacetophenone as a flavor compound in stale dry milk. *Nature* 202:185–187.
28. Karagul-Yuceer, Y., K. N. Vlahovic, M. A. Drake, and K. R. Cadwallader. 2003. Characteristic aroma components of rennet casein. *J Agric Food Chem* 51:6797–6801.
29. Holland, J. F., and B. D. Gardner. 2002. The advantages of GC–TOFMS for flavor and fragrance analysis. In *Flavor, Fragrance and Odor Analysis*, ed. R. Marsili, 107–138. New York: Marcel Dekker, Inc.
30. Serodio, P., and J. M. F. Nogueira. 2005. Development of a stir bar sorptive extraction liquid desorption-large volume injection capillary gas chromatographic-mass spectrometric method for pyrethroid pesticides in water samples. *Anal and Bioanalytical Chem* 382(4):1141–1151.
31. Harmon, A. D. 2002. Solid-phase microextraction for the analysis of aromas and flavors. In *Flavor, Fragrance and Odor Analysis*, ed. R. Marsili, 92–94. New York: Marcel Dekker, Inc.
32. Werkhoff, P., S. Brennecke, W. Bretschneider, and H-J. Betram. 2002. Modern methods for isolating and quantifying volatile flavor and fragrance compounds. In *Flavor, Fragrance and Odor Analysis*, ed. R. Marsili, 179–194. New York: Marcel Dekker, Inc.
33. Andrewes, P., J. G. Bendall, G. Davey, and R. Shingleton. 2010. A musty flavour defect in calcium caseinate due to chemical tainting by 2,4.6-trimbromophenol and 2,4,6-tribromoanisole. *Int Dairy J* 20:423–428.
34. Cant, P. A. E. 1979. Identification of α, α -dibromoanisole as a trace flavour contaminant of an abnormal commercial casein. *New Zealan J Dairy Sci and Tech* 14:35–41.
35. Wihlborg, R., D. Pippitt, and R. Marsili. 2008. Headspace sorptive extraction and GC–TOFMS for the identification of volatile fungal metabolites. *J Microbiol Methods* 75:244–250.
36. Larsen, T. O., and J. C. Frisvad. 1994. A simple method for collection of volatile metabolites from fungi based on diffusive sampling from Petri dishes. *J Microbiol Methods* 19:297–305.

chapter five

The olfactometric analysis of milk volatiles with a novel gas chromatography-based method

A case study in synergistic perception of aroma compounds

Yvette Naudé and Egmont R. Rohwer

Contents

Introduction

Consumers consider the cooked cabbage-like, sulfurous and stale notes imparted by the ultrahigh-temperature (UHT) treatment of packaged long-life milk undesirable. (Contarini et al. 1997; Nursten 1997; Perkins et al. 2005; Simon and Hansen 2001; Vazquez-Landaverde et al. 2005; Vazquez-Landaverde, Torres, and Qian 2006). Analytical methods that have been used to study the aroma of dairy products are gas chromatography-mass spectrometry (GC–MS) and gas chromatography–olfactometry (GCO) (Bendall 2001; D'Acampora Zellner et al. 2008; Friedrich and Acree 1998; Mahajan, Goddik, and Qian 2004; Moio et al. 1994; Moio and Addeo 1998; Moio, Piombino, and Addeo 2000; Qian, Nelson, and Bloomer 2002; Van Aardt et al. 2005). GCO is traditionally used to identify individual odor active gas chromatographic fractions; a human record of aroma perception, in real time, of the GC effluent at the olfactometer outlet. The rapid elution of compounds is problematic in terms of recalling appropriate odor descriptors. GCO requires utmost concentration and can cause nose fatigue. Typically one to two trained evaluators perform the sniffing. However, a group of evaluators is required for reliable GCO analysis, necessitating numerous analyses and multiple gas chromatographs equipped with sniff ports (Van Ruth 2001). Instrumental analyses are not always performed on the aroma-relevant compounds but also on nonodorous compounds that may include hazardous chemicals (Ampuero and Bosset 2003). Coelution of compounds is a common occurrence in separating complex mixtures, thus the evaluator may not realize that composite peaks, instead of pure compounds, are sniffed. Potential synergistic effects cannot be observed where single compounds are evaluated over time. Many synergistic effects are known to occur between single compounds in complex food matrices (Herrmann et al. 2010). Combinations of single substances can produce enhancing or masking interactive effects. Perceived synergistic sensory effects can arise not only from a blend of similar volatiles, but also from a blend of chemically unrelated compounds; for example, the sensory threshold for vanillin is lower in the presence of oak lactones (Perez-Coello, Sanz, and Cabezudo 1997). Interestingly, the end result of combining single substances in complex food matrices may be the emergence of a strikingly different sensory perception, completely unrelated to that of the individual compounds alone. Herrmann et al. (2010) evaluated the aging of beer and reported a change in the sensory perception of components when in combination. Here, E-2-nonenal was described by tasters as cardboard-like when the single substance was evaluated, whereas (E)-2-(Z)-6-nonadienal was described as cucumber-like. However, the combined effect of the two compounds produced a sweet fruity flavor sensory perception distinctly different from the single substances alone. This new sensory perception appears related to both the absolute and relative concentrations of the two compounds. Because the generation of new odors by addition of individual compounds is not yet fully understood, it is hard to determine how individual compounds of a product relate to the perception of its overall aroma when using traditional instrumental techniques (Ampuero and Bosset 2003).

Addressing the limitation of evaluating single compounds as opposed to mixtures, we developed a heart-cut gas chromatographic fraction collection (GCFC) technique to study, off-line, synergistic perceptions of aroma compounds. The target compounds selected for technique evaluation in this study are important thermally derived off-flavor compounds present in UHT milk. Powerful odor active volatiles detected in heated milk include 2-heptanone, 2-nonanone, and nonanal, with 2-heptanone and 2-nonanone identified as the most intense volatile flavor compounds of UHT milk (Moio et al. 1994). Higher concentrations were found of, among others, the sulfur compound dimethyl sulphide; the ketones: 2-hexanone, 2-heptanone, 2-nonanone; and the aldehydes: 2-methyl-propanal, 3-methylbutanal, and decanal in UHT milk when compared with concentrations measured in raw and pasteurized milk (Vazquez-Landaverde et al. 2005). Calculated odor activity values for dimethyl sulphide, 2,3-butanedione, 2-heptanone, 2-nonanone, 2-methylpropanal, 3-methylbutanal, nonanal, and decanal show that these compounds may well be important contributors to the off-flavor of UHT milk (Vazquez-Landaverde et al. 2005). The formation of 2,3-butanedione is, however, not only heat-induced but can also point to microbial activity in milk (Vazquez-Landaverde et al. 2005). We report on the sorptive extraction of aroma compounds using multichannel open tubular silicone (polydimethylsiloxane [PDMS]) rubber traps (MCTs) for quantitative gas chromatographic analysis; for fractionation and collection of aroma compounds from the total chromatogram; for multidimensional GC; and for off-line olfactory evaluation of synergistic effects.

Sorptive extraction

Commercial sample-enrichment techniques

Solvent-free approaches to the isolation of volatile components from food matrices include headspace solid-phase microextraction (HS-SPME), stir bar sorptive extraction (SBSE), headspace solid-phase dynamic extraction (HS-SPDE), headspace trap technology, and solid-phase extraction-thermal desorption (SPE-tD) (Bicchi et al. 2000, 2004; Buettner 2007; Contarini et al. 1997, Contarini and Povola 2002; Fabre, Aubry, and

Guichard 2002; Markes International 2010; Marsili 1999, 2000; Perkins et al. 2005; Qian, Nelson, and Bloomer 2002; Ridgway, Lalljie, and Smith 2006, 2007; Schulz et al. 2007; Van Aardt et al. 2005; Vazquez-Landaverde et al. 2005; Vazquez-Landaverde, Torres, and Qian 2006). Headspace sampling provides solvent-free aroma extracts that are more representative of food aroma when compared with those obtained by solvent extraction. Commercial sorptive extraction techniques, such as SPME, SBSE, SPDE, and headspace trap technology, offer efficient concentration and can be used for sampling the headspace (Bicchi et al. 2000, 2004; Marsili 1999, 2000; Perkins et al. 2005; Ridgway, Lalljie, and Smith 2006, 2007; Schulz et al. 2007; Tienpont et al. 2000). SPME, pioneered by Janusz Pawliszyn and his group, is a fiber coated with a small sorbent volume of 0.6–0.9 µL (Bicchi et al. 2000, 2004; Schulz et al. 2007; Zhang and Pawliszyn 1993), SBSE, introduced by Pat Sandra and his team, is a glass-encapsulated magnetic stir bar coated with volumes of 25–200 µL PDMS (Baltussen et al. 1999; Bicchi et al. 2004), SPDE is a sorptive coating with a sorbent volume of 6 µL on the inside wall of a stainless steel needle of a gastight syringe (Bicchi et al. 2004; Ridgway, Lalljie, and Smith 2006, 2007; Schulz et al. 2007), whereas headspace trap technology makes use of a headspace sampler and a built-in trap (Schulz et al. 2007). The built-in trap is a tube filled with a solid sorbent with a volume of 160 µL (Schulz et al. 2007). A SPE-tD cartridge is a titanium tube coated with PDMS on both the inside (1 µm) and the outside (500 µm) of the tube. Two lengths are available with a total exposed PDMS volume of 29.5 µL for the 6 mm length and 147 µL for the 30 mm length, respectively.

Novel multichannel sample-enrichment devices

Multichannel configuration devices designed for air pollution studies have been utilized by Lane et al. (1988) who coated a collection of glass tubes on the inside and on the outside with stationary phase, and by Kriegler and Hites (1992) who used a bundle of fused silica tubes cut from a single DB-1 capillary GC column. Practical limitations were complicated instrumental arrangements and high pressure drops because of the longer length of the traps (25–60 cm). Ortner and Rohwer (1996) constructed relatively shorter length (10.5–12.5 cm) thick film silicone rubber traps in a novel multichannel configuration to concentrate semivolatile organic air pollutants. The traps were desorbed in an inlet similar to a programmable temperature vaporization injector. McGee and Purzycki (1998) designed the Zenith trap, a bundle of capillary GC tubes, each coated on the inside with polymers of different polarity. The size of the bundle was made to fit into a thermal desorber. The authors collected the scent of a blue hyacinth flower on the Zenith trap.

Multichannel silicone (PDMS) rubber traps

The headspace traps used for the isolation of aroma volatile compounds from UHT milk are MCTs prepared in-house (Naudé, Van Aardt, and Rohwer 2009; Ortner and Rohwer 1996; Rohwer, Lim Ah Tock, and Naudé 2006; Sivakumar et al. 2008). MCTs containing 0.4 ± 0.02 g silicone were prepared based on a technique described by Ortner and Rohwer (1996). The MCT was designed to fit a commercial thermal desorber system (TDS) available from Gerstel. A bundle of 22 channels of silicone elastomer medical grade tubing (0.64 mm OD × 0.3 mm ID, Sil-Tec, Technical Products, Georgia) are inserted into 17.8-cm-long glass desorption tubes (4 mm ID, 6.00 mm OD) (Figure 5.1). The MCT inside the desorption tube is 55-mm long (Figure 5.1). The ends of the glass desorption tube are capped with glass stoppers. The glass stoppers are held in place by Teflon tubing. The MCT with its trapped analytes is not exposed to the Teflon tubing, thereby preventing potential adsorption of analytes onto the Teflon. Compared to commercial sorption devices, the MCT provides a larger sample-enrichment capacity of 600 µL PDMS. MCTs are not limited to desorption tubes from a single manufacturer; they are constructed to fit into any of the commercial tD tubes available. Shorter length MCTs are also made to suit specific applications. In contrast to multichannel traps consisting of a bundle of GC capillary columns which contain nonsorptive outer coatings, the MCT is a bundle of PDMS tubes, thus both the insides and the outsides of the tubing provide sorptive surfaces. Further advantages of the MCT over commercial sorption devices are its open tubular structure and low pressure drop associated with multichannel flow (Ortner and Rohwer 1996). These characteristics are not only particularly suited to the recapturing of chromatographic fractions—single-aroma compounds and their combinations from the GC effluent during a GC run—but also for the off-line release of the heart-cut aroma compounds from the MCTs for olfactometric evaluation. Isolating headspace aroma compounds from beverages onto MCTs and GC fraction collection (heart-cutting) onto MCTs for off-line olfactometry were successfully applied to the aroma studies of beer (Lim Ah Tock 2009) and of novel mocha-styled pinotage wines (Naudé and Rohwer 2010b). Aroma analytes are concentrated from matrices onto MCTs by either a purge-and-trap method, for example when sampling beverages (Lim Ah Tock 2009; Naudé, Van Aardt, and Rohwer 2009; Naudé and Rohwer 2010b; Potgieter 2006) or liquidized fruit (Sivakumar et al. 2008), or by dynamic headspace sampling using an air sampling pump when sampling live plants. Neat compounds, either single or their combinations, are conveniently trapped onto an MCT by simply drawing with a syringe a volume of their headspace through an MCT. After sample enrichment, the aroma is released from the MCTs for off-line olfactometric evaluation. In addition, MCTs containing aroma analytes are thermally desorbed into a gas chromatograph with flame ionization detection (GC–FID) or GC–MS.

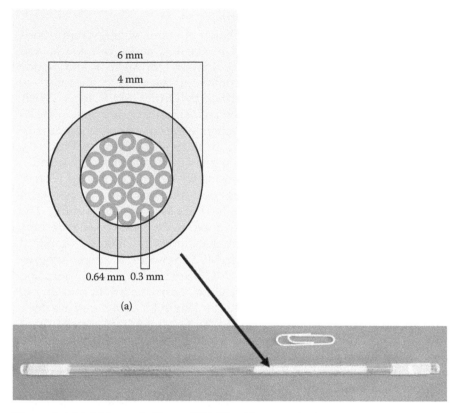

Figure 5.1 Multichannel silicone (PDMS) rubber trap (MCT). (a) Cross section of an MCT. (b) Twenty-two silicone elastomer (PDMS) tubes (sorption volume of 600 µL) are arranged in parallel inside of a commercial glass thermal desorption (tD) tube. The ends of the tD tube containing the MCT are capped with glass stoppers during storage. Teflon tubing keep the glass stoppers in position, samples are not in contact with the Teflon. (Reprinted from *Journal of Chromatography A*, 1216, Naudé, Y., Van Aardt, M., Rohwer, E. R., Multi-channel open tubular traps for headspace sampling, gas chromatographic fraction collection and olfactory assessment of milk volatiles, page 2799, Copyright 2009, with permission from Elsevier.)

High-capacity headspace sorptive extraction of aroma compounds from milk followed by TDS-CIS-GC–FID

Sample enrichment

A purge-and-trap sampling method developed in-house was used to extract the aroma compounds from packaged long-life UHT milk (2% milk fat) by trapping it on an MCT (Naudé, Van Aardt, and Rohwer 2009). The volatiles were isolated from 200 mL milk inside a 500-mL, flat-bottomed glass flask. The flask was immersed for 35 min in a water bath (40°C), and the sample was purged with 500 mL nitrogen gas at 25 mL/min. The purged volatiles were collected on an MCT at 45°C to prevent the condensation of water from the sample onto the MCT.

Chemical standards

A standard stock solution was prepared in methanol (Saarchem UniVAR, Merck Chemicals (Pty) Ltd., Kempton Park, South Africa) containing (1) dimethyl sulphide; (2) 2-methylpropanal; (3) 2,3-butanedione; (4) 3-methylbutanal; (5) 2-hexanone; (6) 2-heptanone; (7) 2-nonanone; (8) nonanal; and (9) decanal to give a concentration of 1 μg/μL. Standard stock solution was added to 200 mL portions of chilled long-life packaged UHT milk (2% milk fat) to give spiking levels of 0.4, 0.8, 4, 8, 16, 40, 70, 85 μg/200 mL milk. The spiked milk portions were thoroughly mixed on a vortex mixer and returned to the refrigerator to equilibrate for 20 min. After the equilibration period, the spiked samples were extracted as described in Section "Sample enrichment."

Instrumentation

The volatiles were thermally desorbed from the MCT using a TDS from Gerstel installed on an Agilent 6890 GC–FID (Chemetrix, Midrand, South Africa) (Naudé, Van Aardt, and Rohwer 2009). Splitless desorption was from 30°C (3 min) at 60°C/min to 210°C (10 min). The desorption flow rate was 50 mL/min (hydrogen gas, Afrox, Gauteng, South Africa) at 47 kPa. The TDS transfer line temperature was 250°C. The volatiles were cryogenically focused using liquid nitrogen at –100°C on a cooled injection system (CIS). The CIS liner contained silicone rubber to improve retention of the extremely volatile dimethyl sulphide. After desorption of the analytes from the MCT, the CIS was heated to 200°C at 5°C/s and the volatiles were analyzed by GC–FID. The GC oven temperature program was –50°C (3 min) at 10°C/min to 120°C (10 min), 10°C/min to 220°C. Hydrogen carrier gas velocity was 33 cm/s (2.3 mL/min), and the column head pressure was 23 kPa in the constant pressure mode. The GC inlet was in the solvent vent mode (to achieve a high-desorption flow rate for trapping onto the CIS) while the purge valve remained closed to give a "splitless" type injection from the CIS. The GC column was a Zebron ZB-1 30 m × 320 μm ID × 1 μm film thickness (Phenomenex, Separations, Randburg, South Africa).

Novel heart-cutting fraction collection GC-based methods

A fraction collection, heart-cutting, enrichment, and two-dimensional GC approach to aroma evaluation of essential oils was described by Sandra et al. (1980). Here, the column effluent was split using an all glass device. For aroma evaluation, one splitter arm led to an FID and the second splitter arm led to the nose. For off-line heart-cutting, the split effluent was collected on a glass capillary microtrap coated with OV-101. Coelution of two or more compounds frequently occurs when separating complex mixtures (Naudé and Rohwer 2010b; Sandra et al. 1980). Using longer GC columns or a different stationary phase is not always ideal (Sandra et al. 1980). Although sophisticated comprehensive GC × GC offers superior resolving power (Naudé and Rohwer 2010b), similar results may be obtained using one-dimensional GC and GC–MS: a heart-cut of a composite peak from one stationary phase is captured followed by off-line rechromatographing the composite heart-cut on a different GC stationary phase (Naudé and Rohwer 2010a,b; Sandra et al. 1980). Sandra et al.'s (1980) excellent article describes reinjection on a different stationary phase by either flushing analytes from the microtrap with solvent or by connecting the microtrap between the GC inlet and column followed by heating of the GC oven. Commercial development in GC sample introduction systems dictated our strategy of conveniently desorbing the MCT in a thermal desorber. A novel multidimensional GC approach is followed where MCTs containing selectively trapped compounds are redesorbed for compound identification by off-line second-dimension separation by GC–MS (Naudé, Van Aardt, and Rohwer 2009) or by the complimentary technique of comprehensive GC × GC–time of flight mass spectrometer (TOFMS) (Naudé and Rohwer 2010a,b). Recapturing of odor-active compounds from the GC effluent onto secondary MCTs for analysis by GC–MS is especially useful when dealing with compounds of unknown identity (Naudé, Van Aardt, and Rohwer 2009; Naudé and Rohwer 2010b) (Figure 5.2). The multidimensional GCFC and off-line olfactometric technique is particularly suited to a directed search approach when confronted with complex aroma chromatograms: heart-cuts of composite peaks obtained from a first-dimension apolar separation can be reinjected for a second-dimension polar separation followed yet again by the selective heart-cutting of single peaks or their combinations for off-line olfactometric evaluation.

Capturing of single compounds and their combinations by off-line heart-cut GCFC

A chromatographic profile of the packaged long-life UHT milk is first obtained using the conditions described in *Instrumentation*. Integration

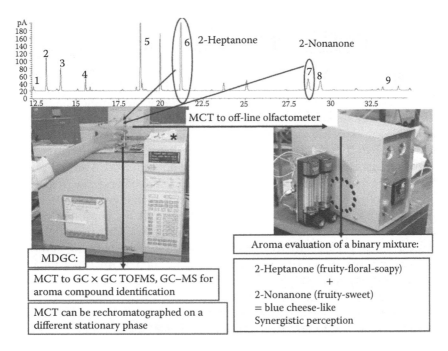

Figure 5.2 Fraction collection and off-line evaluation of synergistic perceptions of aroma compounds. (1) Dimethyl sulphide; (2) 2-methylpropanal; (3) 2,3-butanedione; (4) 3-methylbutanal; (5) 2-hexanone; (6) 2-heptanone; (7) 2-nonanone; (8) nonanal; and (9) decanal. (*) Collector removed from flame ionization detection (FID). FID conditions: detector 300°C, makeup gas (He) 50 mL/min, detector gases off. The MCT is simply placed on top of the FID flame tip and held in position during peak elution. The MCT is removed from the flame tip at the end of peak elution. Peaks 6 and 7 were both selectively captured on the same MCT. The combined aroma (6 + 7) mixture was released from the MCT at 160°C with a flow of N_2 gas (20 mL/min). The synergistic perception of the combined aroma (6 + 7) mixture was evaluated by off-line olfactometry. The portable olfactometric device consists of two sniff ports for a blank and a sample, gas flow control and temperature programming. Multidimensional GC: MCTs containing fractions can be rechromatographed on a different stationary phase and by GC–MS for compound identification. (Chromatogram reprinted from *Journal of Chromatography A*, 1216, Naudé, Y., Van Aardt, M., Rohwer, E. R., Multichannel open tubular traps for headspace sampling, gas chromatographic fraction collection and olfactory assessment of milk volatiles, page 2802, Copyright 2009, with permission from Elsevier.)

results of peak start time and peak end time of the chromatogram of the UHT milk are then used to establish the heart-cutting event times. In a subsequent run, various sections of the GC effluent are selectively recaptured onto secondary MCTs on a carefully timed basis (Figure 5.2). Manual collection of heart-cut peaks from the complex/total chromatogram commences, as a rule of thumb, 30 s before peak elution and ends 10 s after elution. For GCFC, the instrumental conditions were as previously described; the only difference now was the modification of the detector parameters. The electrometer is switched off. The detector top assembly and the collector are easily removed by loosening the knurled brass nut of the detector assembly and by lifting the collector out. Single peaks and their combinations are collected at the end of the GC column by simply placing a secondary MCT on the inactivated FID flame tip and by support-ing the MCT in this position by hand (Figure 5.2). Complex instrumental setups, sophisticated equipment, or valves are not required. The MCT was placed on the flame tip before peak elution and removed once the peak had eluted. To collect single peaks, nine MCTs were exchanged one after the other so that each MCT contained a single compound. For combinations of compounds, two peaks were collected on a single MCT. During collection, the FID and flame gases (hydrogen and air) were switched off, the makeup gas (helium) plus carrier gas (hydrogen) flow totalled 50 mL/min and the detector temperature was at 300°C. The low pressure drop of the open tubular channels of the MCT allows most of the effluent of the FID to pass through the trap without special sealing arrangement that would other-wise complicate the GCFC procedure. The advantage of a low pressure drop is not offered by conventional packed traps. After single compounds or their combinations were selectively recaptured, the MCTs were capped with custom-made glass and Teflon stoppers (Figure 5.1) and stored for olfactory evaluation or for GC–MS identification.

Off-line olfactory evaluation of single compounds and their combinations

Nine odor-active compounds (sulphide, aldehydes, and ketones) were selected for evaluation and comparison with the aroma of the overall milk profile. The aroma was released in a controlled manner by heating the sec-ondary MCT containing the recaptured component/s from the GC efflu-ent in a portable heating device with a flow of nitrogen gas at a flow rate of 20 mL/min (Figure 5.2). The portable olfactometric device was designed and built in-house. The MCTs were inserted in the cavity in the portable heating device (Figure 5.2). The compounds were released from the sec-ondary MCTs at GC column elution temperature for the olfactory assess-ment of the single compounds (Table 5.1). As the aroma eluted from the MCT, it was sniffed by a team consisting of six experienced nonsmoking evaluators (five female, one male). All six of the evaluators participated as a group during the same assessment session. For the evaluation of com-binations of single compounds, the combined fractions were released at 130°C or at 160°C, and the aroma was evaluated by a panel of three per-sons. A second device since built contained two sniff ports so that a blank

Table 5.1 Olfactory Assessment of Single Compounds Captured by GCFC on Individual Secondary MCTs,[a] the Overall Milk Aroma Collected on a Primary MCT[b] and Binary Combinations of Compounds Captured by GCFC onto Secondary MCTs[c]

1. Single compounds captured by GCFC onto secondary MCTs

Single compounds GCFC	Aroma release temperature (°C)	[a]Aroma description for single compounds (n = 6)		
		Spiked UHT milk 195–245 µg/L	Unspiked UHT milk 1–44 µg/L	
1. Dimethyl sulphide	21	Sulfurous, rotten egg+++[d]	–	
2. 2-Methylpropanal	50	Malty, chemical sweet+++	–	
3. 2,3-Butanedione	60	Butter+++	–	
4. 3-Methylbutanal	73	Malty, chemical sweet+++	Chemical sweet+	
5. 2-Hexanone	103	Feed, hay, field, veld grass++	–	
6. 2-Heptanone	128	Fragrant soap, fruity, floral++	Fragrant soap+	
7. 2-Nonanone	178	Fruity, sweet+++	Fruity, sweet++	
8. Nonanal	185	Biscuit, cotton candy, caramel++	–	
9. Decanal	194	Dusty, musty, rubber++	Dusty, rubber+	

2. Whole milk aroma profile captured by purge-and-trap sampling onto a primary MCT (nonfractionated aroma profile)

	Aroma release temperature (°C)	[b]Aroma description for the overall milk aroma (n = 3)	
		Spiked UHT milk 195–245 µg/L	Unspiked UHT milk 1–44 µg/L
	24–40	Rotten egg+, butter+++, sweet+, malty+++	–
	60–120	Malty+, feed, pungent, sour milk, sweaty+++	Chemical sweet+
	120–185	Sour milk, burnt milk++	Fragrant soap, floral+

(Continued)

Table 5.1 Olfactory Assessment of Single Compounds Captured by GCFC on Individual Secondary MCTs,[a] the Overall Milk Aroma Collected on a Primary MCT[b] and Binary Combinations of Compounds Captured by GCFC onto Secondary MCTs[c] (*Continued*)

3. Combinations of compounds captured by GCFC onto secondary MCTs

Binary mixtures	Aroma release temperature (°C)	[c]Aroma description for combinations of compound (n = 3) Spiked UHT milk 195–245 µg/L
4. 3-Methylbutanal 5. 2-Hexanone	130	Malty+++, sweet+
5. 2-Hexanone 6. 2-Heptanone	160	Milky-sweet, condensed milk++
6. 2-Heptanone 7. 2-Nonanone	160	Feta cheese, blue cheese, sweaty, sour, fatty+++
5. 2-Hexanone 7. 2-Nonanone	160	Cheesy, sweaty, sour+

Source: Reprinted from *Journal of Chromatography A*, 1216, Naudé, Y., Van Aardt, M., Rohwer, E. R., Multi-channel open tubular traps for headspace sampling, gas chromatographic fraction collection and olfactory assessment of milk volatiles, page 2803, Copyright 2009, with permission of Elsevier.

GCFC = gas chromatographic fraction collection; PDMS = polydimethylsiloxane; MCTs = multichannel open tubular silicone (PDMS) rubber traps; UHT = ultrahigh temperature.

[a] Olfactory evaluation of packaged long-life UHT milk (2% milk fat): spiked and unspiked. For each sample nine secondary MCTs each containing heart-cuts of single compounds were individually desorbed in the portable sniffing device at specific temperatures with a flow of nitrogen gas of 20 mL/min.

[b] Olfactory evaluation of the purge-and-trap extracted overall milk aroma (primary MCT, no GCFC) of long-life packaged UHT milk (2% milk fat): spiked and unspiked. The MCTs were desorbed in the portable sniffing device from 24°C to 185°C over a period of 60 min with a flow of nitrogen gas of 20 mL/min.

[c] Olfactory evaluation of long-life packaged UHT milk (2% milk fat): spiked. For each milk sample four secondary MCTs each containing heart-cuts of binary mixtures of compounds were individually desorbed in the portable sniffing device with a flow of nitrogen gas of 20 mL/min.

[d] Intensity indicator.

MCT can be sniffed next to an MCT containing the trapped aroma compounds (Naudé and Rohwer 2010b).

Slow release of the aroma of single compounds captured by GCFC on individual secondary MCTs

Results of the olfactory assessment of the individual compounds captured from UHT milk (spike levels of 195–245 µg/L) are given in Table 5.1. The aroma descriptors reported by the evaluators for each of the individual compounds generally matched the descriptors reported in the literature (Nursten 1997; Friedrich and Acree 1998; Rychlik, Schieberle, and Gosch 1998). A fairly neutral background was described for the recaptured single compounds collected from the chromatogram of the unspiked UHT milk (Table 5.1).

Slow release of the total milk aroma profile captured by purge-and-trap on primary MCTs

The aroma from primary MCTs containing the total milk profile extracted by purge-and-trap was also released for comparison with that of the single compounds and their combinations captured by GCFC (Table 5.1). The MCT in the olfactometric device was heated from ambient to 185°C at 2°C/min. A fairly neutral background was described for the overall milk profile isolated by purge-and-trap from the unspiked (1–44 µg/L) UHT milk (Table 5.1). However, when the overall aroma from the spiked UHT milk (195–245 µg/L) isolated on a primary MCT by purge-and-trap was released, the aroma matched the descriptors for the single compounds up to and including 2-hexanone (heart-cut 5). Thereafter, the aroma observed deviated markedly from the descriptors for the single compounds. Instead of observing the expected aroma of "soapy" (2-heptanone [heart-cut 6]) and "fruity" (2-nonanone [heart-cut 7]), a pungent, sour milk-like, sweaty aroma was noted. This unpleasant aroma was absent when the single compounds recaptured during GCFC were sniffed, and it was absent when the overall aroma profile isolated from the unspiked milk on an MCT was sniffed (Table 5.1). This unexpected and entirely new aroma isolated from the spiked milk on an MCT suggested possible synergistic effects between the compounds.

Slow release of heart-cuts of combinations of single compounds captured by GCFC on individual secondary MCTs

Combinations of the single volatiles were selectively collected onto secondary MCTs during GCFC (Figure 5.2) to determine which compounds were responsible for the change in aroma perception from the expected fruity, soapy, and sweet aroma into the unexpected and wholly new aroma described as pungent, sour milk-like, and sweaty (Table 5.1). For each of the binary sets, the combinations were collected on a single MCT (four MCTs, each contained a heart-cut binary set). For example, heart-cuts 4 + 5 (3-methylbutanal + 2-hexanone) were both collected on a single MCT. The MCT was simply placed on the flame tip of the inactive FID just before the elution of 3-methylbutanal. The MCT was removed from the flame tip once the peak eluted. The same MCT containing 3-methylbutanal was once again placed on the flame tip just before the elution of 2-hexanone. The MCT was removed from the flame tip once the peak eluted. The MCT containing the two heart-cuts was capped for storage. The portable sniff device was set to a specific temperature, the MCT containing both compounds was inserted, and the combined aroma was released with a flow of nitrogen. A malty aroma dominated when a combination of the aldehyde 3-methylbutanal (malty, sweet), and the ketone 2-hexanone (feed) was sniffed (heart-cuts 4 + 5) (Table 5.1). A milky-sweet, condensed milk aroma was described for the combination of the ketones, 2-hexanone (feed) and 2-heptanone (soapy, fruity, floral) (heart-cuts 5 + 6). A pungent, feta cheese, blue cheese, sweaty, sour, fatty aroma was observed when a combination of the ketones, 2-heptanone (soapy, fruity, floral) and 2-nonanone (fruity, sweet), was sniffed (heart-cuts 6 + 7) (Table 5.1). A low intensity cheesy, sweaty, sour aroma was noted when a combination of the ketones, 2-hexanone (feed) and 2-nonanone (fruity, sweet), was sniffed (heart-cuts 5 + 7) (Table 5.1).

The results from the olfactory assessment show that the aroma of the individual ketones is distinctly different to the aroma of mixtures of ketones. The combination of 2-nonanone with either 2-heptanone or 2-hexanone, gave a cheesy odor, whereas the combination of 2-hexanone and 2-heptanone did not give a cheesy aroma. The combination of 2-heptanone and 2-nonanone was responsible for the pungent, sweaty, and sour milk-like note observed when the overall milk aroma captured by purge-and-trap was sniffed. A strong blue cheese-like note was detected for a combination of 2-heptanone and 2-nonanone selectively captured from the total chromatogram by GCFC. These results seem consistent with the impact that 2-heptanone and 2-nonanone has on the aroma of UHT milk and of blue-veined cheese. Moio et al. (1994) applied CharmAnalysis to identify 2-heptanone and 2-nonanone as the two most intense flavor compounds likely to be the main odorants of UHT milk. Vazquez-Landaverde et al. (2005) reported that the odor activity values for 2-heptanone and 2-nonanone suggest that these compounds could be very important contributors to the aroma of heated milk. Moio, Piombino, and Addeo (2000) concluded that 2-heptanone and 2-nonanone impart a strong Gorgonzola (Italian blue-veined cheese) cheese note in their study of odor-impact compounds of Gorgonzola cheese. Patton (1950) and Qian,

Nelson, and Bloomer (2002) reported that compounds central to the blue cheese aroma include 2-heptanone and 2-nonanone, and McGorrin (2002) stated that 2-heptanone and 2-nonanone are reportedly the dominant character compounds of blue cheese flavor.

Summary

Synergistic sensory effects are known to emerge between single compounds in complex food matrices. To overcome the limitations of traditional GCO where single compounds are evaluated over time not providing information on their behavior in mixtures—we developed a heart-cut GCFC technique to study, off-line, the synergistic perceptions of aroma compounds. Single compounds and their combinations are selectively recaptured from the GC effluent onto an MCT without the need for complicated instrumental arrangements. A benefit of the MCT is its low pressure drop not offered by conventional packed traps. The aroma heart-cuts are released off-line in a controlled manner by heating the MCT in a portable device. An advantage of uncoupling the time scale is that the olfactory assessment can be terminated and resumed at convenience when the evaluators indicate nose fatigue. Olfactory results suggest that a synergistic combination of 2-heptanone and 2-nonanone was responsible for a pungent cheese, sour milk-like aroma detected in UHT milk.

Acknowledgments

We thank Dr. Fanie van der Walt for constructing the olfactory device, and the olfactory panel: Dr. Marleen van Aardt, Professor Elna Buys, Dr. Henriette de Kock, Marise Kinnear, and Alex Zabbia of the Department of Food Science, University of Pretoria, Pretoria, South Africa. We are grateful to Sasol Fuels and the South African National Research Foundation (NRF) for funding.

References

Ampuero, S. and J. O. Bosset. 2003. The electronic nose applied to dairy products: a review. *Sensors Actuat B* 94:1–12.

Baltussen, E., P. Sandra, F. David, and C. Cramers. 1999. Stir bar sorptive extraction (SBSE), a novel extraction technique for aqueous samples: Theory and principles. *J Microcol Separ* 11:737–747.

Bendall, J. G. 2001. Aroma compounds of fresh milk from New Zealand cows fed different diets. *J Agric Food Chem* 49:4825–4832.

Bicchi, C., C. Cordero, C. Iori, P. Rubiolo, and P. Sandra. 2000. Headspace sorptive extraction (HSSE) in the headspace analysis of aromatic and medicinal plants. *J High Resol Chromatogr* 23(9):539–546.

Bicchi, C., C. Cordero, E. Liberto, P. Rubiolo, and B. Sgorbini. 2004. Automated headspace solid-phase dynamic extraction to analyse the volatile fraction of food matrices. *J Chromatogr A* 1024:217–226.

Buettner, A. 2007. A selective and sensitive approach to characterize odour-active and volatile consitituents in small-scale human milk samples. *Flavour Frag J* 22:465–473.

Contarini, G. and M. Povola. 2002. Volatile fraction of milk: Comparison between purge and trap and solid phase microextraction techniques. *J Agric Food Chem* 50:7350–7355.

Contarini, G., M. Povolo, R. Leardi, and P. M. Toppino. 1997. Influence of heat treatment on the volatile compounds of milk. *J Agric Food Chem* 45:3171–3177.

D'Acampora Zellner, B., P. Dugo, G. Dugo, and L. Mondello. 2008. Gas chromatography-olfactometry in food flavour analysis. *J Chromatogr A* 1186: 123–143.

Fabre, M., V. Aubry, and E. Guichard. 2002. Comparison of different methods: Static and dynamic headspace and solid-phase microextraction for the measurement of interactions between milk proteins and flavor compounds with an application to emulsions. *J Agric Food Chem* 50:1497–1501.

Friedrich, J. E. and T. E. Acree. 1998. Gas chromatography olfactometry (GC/O) of dairy products. *Int Dairy J* 8(3):235–241.

Herrmann, M., B. Klotzbücher, M. Wurzbzcher et al. 2010. A new validation of relevant substances for the evaluation of beer aging depending on the employed boiling system. *J Inst Brewing* 116(1):41–48.

Kriegler, M. S. and R. A. Hites. 1992. Diffusion denuder for the collection of semivolatile organic compounds. *Environ Sci Tech* 26:1551–1555.

Lane, D. A., N. D. Johnson, S. C. Barton, G. H. S. Thomas, and W. H. Schroeder. 1988. Development and evaluation of a novel gas and particle sampler for semivolatile chlorinated organic compounds in ambient air. *Environ Sci Tech* 22(8):941–947.

Lim Ah Tock, M. J. 2009. Aroma analysis of alcoholic beverages using multichannel silicone rubber traps. M.Tech. dissertation. Tshwane University of Technology, Pretoria, South Africa.

Mahajan, S. S., L. Goddik, and M. C. Qian. 2004. Aroma compounds in sweet whey powder. *J Diary Sci* 87:4057–4063.

Markes International Thermal Desorption Technical Support Note 88. 2010. Enhancing olfactory profiling of fruit juices and wine using complementary analytical thermal desorption techniques.

Markes International Thermal Desorption Technical Support Note 94. 2010. Using Markes' thermal desorption technology to automate high/low analysis of complex beer sample.

Marsili, R. T. 1999. SPME-MS-MVA as an electronic nose for the study of off-flavors in milk. *J Agric Food Chem* 47:648–654.

Marsili, R. T. 2000. Shelf-life prediction of processed milk by solid-phase microextraction, mass spectrometry, and multivariate analysis. *J Agric Food Chem* 48:3470–3475.

McGee, T., and K. L. Purzycki. 2002. Headspace techniques for the reconstitution of flower scents and indentification of new aroma chemicals, in: *Flavor, Fragrance and Odor Analysis*, ed. R. Marsili, 249–276. New York, NY: Marcl Dekker, Inc.

McGorrin, R. J. 2002. Character impact compounds: flavors and off-flavors in foods, in: *Flavor, Fragrance and Odor Analysis*, ed. R. Marsili, 375–413. New York, NY: Marcel Dekker, Inc.

Moio, L., P. Etievant, D. Langlois, J. Dekimpe, and F. Addeo. 1994. Detection of powerful odorants in heated milk by use of extract dilution sniffing analysis. *J Dairy Res* 61:385–395.

Moio, L., and F. Addeo. 1998. Grana Padano cheese aroma. *J Dairy Res* 65:317–333.

Moio, L., P. Piombino, and F. Addeo. 2000. Odour impact compounds of Gorgonzola cheese. *J Dairy Res* 67:273–285.

Naudé, Y., and E. R. Rohwer. 2010a. New multidimensional chromatographic methods for enantiomeric analysis of *o,p'*-DDT and associated environmental pollutants, presented at the 34th *International Symposium on Capillary Chromatography*, Riva del Garda, Italy, May 30–June 4, 2010.

Naudé, Y., and E. R. Rohwer. 2010b. The analysis of pinotage wine aromas by a novel fractionating olfactometric method combined with comprehensive two dimensional gas chromatography—mass spectrometry, presented at the 15th World Congress of Food Science and Technology, Cape Town, South Africa, August 22–26, 2010.

Naudé, Y., M. Van Aardt, and E. R. Rohwer. 2009. Multi-channel open tubular traps for headspace sampling, gas chromatographic fraction collection and olfactory assessment of milk volatiles. *J Chromatogr A* 1216:2798–2804.

Nursten, H. E. 1997. The flavour of milk and dairy products: I. Milk of different kinds, milk powder, butter and cream. *Int J Dairy Tech* 50(2):48–56.

Ortner, E. K., and E. R. Rohwer. 1996. Trace analysis of semi-volatile organic air pollutants using thick film silicone rubber traps with capillary gas chromatography. *J High Resol Chromatogr* 19:339–344.

Patton, S. 1950. The methyl ketones of blue cheese and their relation to its flavor. *J Dairy Sci* 33(9):680–684.

Perkins, M. L., B. R. D'Arcy, A. T. Lisle, and H. C. Deeth. 2005. Solid phase microextraction of stale flavour volatiles from the headspace of UHT milk. *J Sci Food Agric* 85:2421–2428.

Perez-Coello, M. S., J. Sanz, and M. D. Cabezudo. 1997. Analysis of volatile components of oak wood by solvent extraction and direct thermal desorption-gas chromatography-mass spectrometry. *J Chromatogr A* 778:427–434.

Potgieter, N. 2006. Analysis of beer aroma using purge-and-trap sampling and gas chromatography. M.Sc. dissertation. University of Pretoria, Pretoria, South Africa.

Qian, M., C. Nelson, and S. Bloomer. 2002. Evaluation of fat-derived aroma compounds in blue cheese by dynamic headspace GC/Olfactometry-MS. *J Am Oil Chem Soc* 79:663–667.

Ridgway, K., S. P. D. Lalljie, and R. M. Smith. 2006. Comparison of in-tube sorptive extraction techniques for non-polar volatile organic compounds by gas chromatography with mass spectrometric detection. *J Chromatogr A* 1124:181–186.

Ridgway, K., S. P. D. Lalljie, and R. M. Smith. 2007. Use of in-tube sorptive extraction techniques for determination of benzene, toluene, ethylbenzene and xylene in soft drinks. *J Chromatogr A* 1174:20–26.

Rohwer, E. R., M. J. Lim Ah Tock, and Y. Naudé. 2006. SA provisional patent application ZA 2006/07538.

Rychlik, M., P. Schieberle, and W. Gosch. 1998. Compilation of odor thresholds, odor qualities and retention indices of key food odorants. Deutsche Forschungsanstalt für Lebensmittelchemie and Institut für Lebensmittelchemie der Technischen Universität München, Garching, Germany, ISBN 3-9803426-5-4.

Sandra, P., T. Saeed, G. Redant, M. Godefroot, M. Verstappe, and M. Verzele. 1980. Odour evaluation, fraction collection and preparative scale separation with glass capillary columns. *J High Resol Chromatogr Comm* 3:107–114.

Schulz, K., J. Drebler, E. Sohnius, and D. W. Lachenmeier. 2007. Determination of volatile constituents in spirits using headspace trap technology. *J Chromatogr A* 1145:204–226.

Simon, M., and A. P. Hansen. 2001. Effect of various dairy packaging materials on the shelf life and flavor of ultrapasteurized milk. *J Dairy Sci* 84:784–791.

Sivakumar, D., Y. Naudé, E. Rohwer, and L. Korsten. 2008. Volatile compounds, quality attributes, mineral composition and pericarp structure of South African litchi export cultivars Mauritius and McLean's Red. *J Sci Food Agric* 88:1074–1081.

Tienpont, B., F. David, C. Bicchi, and P. Sandra. 2000. High capacity headspace sorptive extraction. *J Microcol Separ* 12(11):577–584.

Van Aardt, M., S. E. Duncan, J. E. Marcy, T. E. Long, S. F. O' Keefe, and S. R. Nielsen-Sims. 2005. Aroma analysis of light-exposed milk stored with and without natural and synthetic antioxidants. *J Dairy Sci* 88:881–890.

Van Ruth, S. M. 2001. Methods for gas chromatography-olfactometry: a review. *Biomol Eng* 17:121–128.

Vazquez-Landaverde, P. A., G. Velazquez, J. A. Torres, and M. C. Qian. 2005. Quantitative determination of thermally derived off-flavor compounds in milk using solid-phase microextraction and gas chromatography. *J Dairy Sci* 88:3764–3772.

Vazquez-Landaverde, P. A., J. A. Torres, M. C. J. Qian. 2006. Effect of high-pressure-moderate-temperature processing on the volatile profile of milk. *J Agric Food Chem* 54:9184–9192.

Zhang, Z. and J. Pawliszyn. 1993. Headspace solid-phase microextraction. *Analyt Chem* 65:1843–1852.

chapter six

Characterizing aroma-active volatile compounds of tropical fruits

Patricio R. Lozano

Contents

The need for diversification of flavor profiles in foods and the newer discoveries about the role of active ingredients in fruits have progressively increased the food industry's interest to study tropical and subtropical fruits in more detail. The term "tropical fruit" refers to a group of fruits that are originally found or massively cultivated in the tropics. This region, characterized by warm temperatures and humid conditions, goes along the equatorial line of the planet and is bordered up north by the Tropic of Cancer and in the south by the Tropic of Capricorn. Nevertheless, this description can get short for fruits that are cultivated at low altitudes in northern regions. For this chapter, we will use the term tropical fruits to describe the fruits that come originally or are produced in the tropical region of the planet and do not tolerate frost.

Flavor is the most important attribute that affects consumption of tropical fruits. It can be considered a combination of senses: predominantly aroma, followed by taste, chemesthesis, texture, and even appearance. Aroma is contributed by a group of aroma-active chemicals with concentrations above that of human detection (odor threshold). These compounds are released from the fruit matrix during mastication, enzymatic activity, or temperature increase. After their release from the fruit matrix, these chemicals reach the olfactory receptors (orthonasally or retronasally) and are later cognitively processed at the brain to associate the flavor sensation fruit (i.e., fruity).

From the aroma-active compounds, a better understanding of the overall flavor profile of a fruit can be obtained. However, it is important to recognize that nonvolatile flavor-active compounds (fatty acids, lipids, sugars, and amino acids) and aroma compounds bound as glycosides in fruits and vegetables also affect the final flavor perception. A complete understanding of the aroma and taste interactions as well as the release of aroma-active compounds bound in tropical fruits is too broad to be covered in a chapter. Consequently, we discuss the identification of the characterizing aroma-active compounds of tropical fruits with high economic potential.

Tropical fruits and their health benefits

Increasing consumption of fruits in recent years is significantly attributed to their health benefits such as their antioxidant properties, their capacity to lower the content of toxic materials or to deplete their effects in the organism.

Tropical fruits are abundant in antioxidants, such as polyphenols, most of which are flavonoids present in ester and glycoside forms (Fleuriet and Macheix 2003), vitamins C, A, B, and E, and carotenoids (Fernandes et al. 2010). For instance, guava (218–270 mg/100 g) and star fruit (278 mg/100 g) have higher antioxidant content expressed as ascorbic acid equivalent antioxidant capacity (AEAC) than that in orange (142 mg/100 g). Antioxidants are related to lowering the incidence of degenerative diseases such as cancer, arteriosclerosis, heart disease, inflammation, and even brain dysfunction. Moreover, they have been claimed to delay aging (Halliwell et al. 1995). Nevertheless, maturation stage of fruits can significantly lower their antioxidant capacity and affect their flavor profile.

The ability to act as a radical scavenger (antiradical capacity), that is, to lower the concentration of oxidizing elements that can affect the human body, is also high in tropical fruits. For instance, acerola contains a higher antiradical capacity than α-tocopherol; acai and cashew apple among others show a higher antiradical capacity than butylated hydroxytoluene, which is the most used commercial antioxidant (Rufino et al. 2009).

In addition to higher amounts of beneficial active ingredients, tropical fruits can provide a wide range of flavors from green-sweet flavors to sulfury-creamy ones and be substantially different compared with orchard fruits. Regardless of their tremendous health benefits, low cost, and distinct flavor profiles, the major constraint to the exportation of tropical fruits to industrialized countries is their common infestation by fruit flies (Tephritidae: Diptera) (Fernandes et al. 2010), which is considered a quarantine pest in the destination countries.

Understanding the main aroma-active compounds of tropical fruits, their formation, and trends during ripening can help identify preservation methods (drying or concentration) that minimize the denaturation or damage of the flavor-characterizing compounds and active ingredients of these fruits.

Flavor characterization of tropical fruits: Free volatiles and glycosidically bound volatiles—A dynamic scenario

Flavor characterization in fruits has passed from identification of a list of compounds obtained after chromatographic analysis of an extract, with little information about their impact on the overall flavor, to monitoring concentrations of odor-impact compounds for specific cultivars of a fruit during controlled maturity stages.

Historically, emphasis has been placed on determining what constitutes flavor rather than on the mechanisms of flavor formation in fruits (Reineccius 2005). However more than often, temperature, enzymatic activity as well as maturity stage of a fruit influence the development of flavor compounds and consequently the reliability and accuracy of a single group of flavor compounds (Laohakunjit and Kerdchoechuen 2007). For instance, enzymes are responsible for the conversion of carbohydrates, lipids, proteins, and amino acids to simpler units and volatile compounds. The rate of flavor formation during fruit ripening is not constant, and it reaches a maximum during the postclimacteric ripening phase (Reineccius 2005). In addition to the continuous change of free volatile compounds in tropical fruits due to enzymatic activity, some flavor compounds are bound as glycosides in fruits. Glycosidically bound flavor compounds typically exist as glucosides, diglycosides, and triglycosides. They are composed of a flavor moiety bound to glucose with the remainder of the glycoside being composed of a number of other simple sugars such as fructose, glactose, and xylose to name a few. The variability and ratio of free and bound volatile compounds in some tropical fruits depends on the cultivar, maturity stage of the fruit, part of the fruit analyzed, and processing conditions. Contribution of glycosidically bound aroma compounds to the overall flavor of a fruit is less than that of the free volatiles because the former group of chemicals needs to be released through physical or chemical pathways.

Isolation of aroma chemicals

Aroma components of the flavor of tropical fruits are mainly analyzed by gas chromatography–mass spectroscopy (GC–MS), whereas their nonvolatile compounds are analyzed by high-pressure liquid chromatography (HPLC). We will concentrate on free aroma volatiles, since their contribution to the overall flavor in most cases is greater than glycosidically bound volatiles one.

In order to study the aroma active chemical compounds in tropical fruits, isolation and concentration of these chemical components distributed in the fruit matrix at low concentrations are needed. Thus, sample preparation needs to consider possible changes due to enzyme activity as well as the cellular disruption that can easily alter the flavor profile. Uses of calcium chloride or liquid nitrogen during grinding a fruit to slow the enzymatic activity are common practices.

Once a sample that represents the flavor profile of the fruit of interest has been prepared, isolation of the volatile compounds is necessary. Most of the techniques used in aroma isolation take advantage of either solubility or volatility of the aroma compounds (Reineccius 2005). On the basis of the principle of isolation of aroma compounds, the methods to obtain the volatile compounds in foods can be broadly grouped into solvent extraction methods, steam distillation methods, headspace techniques, and sorptive techniques (Da Costa and Sanja 2005).

Extraction methods are based on the transfer of the aroma chemicals from the fruit matrix into an organic solvent or a supercritical fluid (carbon dioxide, nitrogen) through agitation, mixing, or centrifugation.

Distillation methods involve the volatilization of the aroma chemicals and posterior condensation and collection with an organic solvent. The main advantage of this method is that the resulting extracts do not contain any nonvolatiles. On the other hand, thermal degradation and highly polar chemicals are extracted poorly. Use of vacuum during distillation of aroma chemicals reduces the chances of thermal degradation and breakdown products.

Headspace collection techniques analyze the aroma compounds that exist in the space above a sample. Although they are simple and quick, and do not involve the use of solvents, the relative concentration of components in headspace does not reflect the concentration in the sample, and they are dependent on the vapor pressure and temperature of the analysis.

Extraction of volatile compounds from a sample using sorption techniques is based on the partitioning of these volatile compounds and a polymeric film. This film can be coated on a needle in the case of SPME or on the outside layer of a magnet stir bar in the case of Twister. It is quick and easy to use and avoid the use of solvent. Nevertheless, sampling conditions as well as the type of coating material and fiber length affect the identification of compounds. Two recently introduced techniques are headspace sorptive extraction (HSSE) and stir bar sorptive extraction (SBSE), which are commercially available as Twister, and are gaining attention in food and flavor analytical laboratories. Twister technology is based on a glass-jacketed stirrer bar coated with adsorbent polymer (polydimethylsiloxane [PDMS]), which can be used to measure the headspace of a sample in a close container or immersed in a liquid sample. This process is less time-consuming and solventless as well as offers higher sensitivity than solid-phase microextraction (SPME). However, owing to the nonpolar nature of the polymer, the extraction has discrimination for polar compounds. (Da Costa and Sanja 2005).

Figure 6.1 illustrates a chromatogram of a combined HSBE and SBSE from a commercial passion fruit concentrate. Simultaneous desorption was conducted using a Gerstel TDU unit coupled to a cryogenic CIS 4

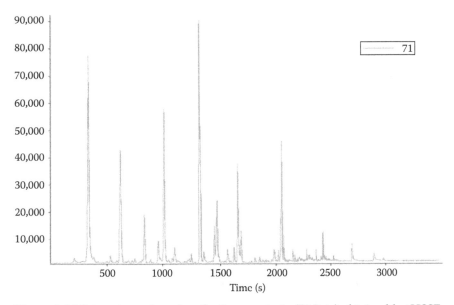

Figure 6.1 Main esters of passion fruit concentrate (50 Brix) obtained by HSSE-SBSE (40°C/30 min).

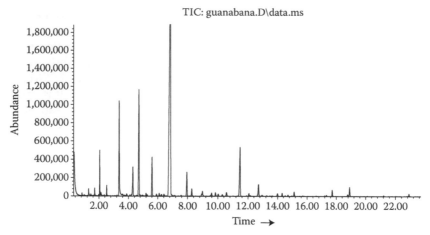

Figure 6.2 Total ion chromatogram of concentrated extract obtained by MASE of soursop concentrate.

injector over a GC–time-of-flight MS (GC–TOFMS) system. For illustration purposes, only mass (m/z 71) characteristic of fruity esters is shown.

Another novel technique with potential application for characterizing fruit essences and water-based samples is membrane-assisted solvent extraction (MASE). This technique is based on a low-density polyethylene membrane (LDPE) of a 0.03-mm wall thickness, which separates the aqueous sample from a nonpolar organic solvent. Generally, an experiment for this technique consists of a glass vial with 15 mL of the aqueous phase and approximately 0.5 mL of hexane. Agitation and temperature allow the compounds of the sample to be transferred to the solvent system. The main disadvantages of this technique are solvent selection to avoid its diffusion through the membrane and need for large injection volumes to identify trace compounds.

The solvent can be injected using a large-volume injection system (LVI); however, concentration under liquid nitrogen can also allow the use of a smaller amount of the sample. Figure 6.2 shows the total ion chromatogram of soursop concentrate (40 Brix) extracted by MASE with solvent concentration to 100 μL. Even though this technique was successful to provide with a representative amount of characterizing flavor compounds for this fruit after concentration, trace compounds previously reported in soursop were not found.

Contribution of aroma compounds to flavor

Natural products usually contain complex mixtures of hundreds of individual compounds. However, the contribution of each of these compounds

varies greatly. Compounds present in trace amounts can be more important contributors than those present in higher concentrations (Da Costa and
Sanja 2005). For instance, sulfur flavor such as (3-methythio-hexyl butyrate)
in guava and ethyl 3-mercaptobutyrate in mango are characterizing high
aroma-impact compounds although their integrated area peaks are not predominant in the chromatogram of these fruit extracts.

Coupling GC with an olfactometry port (GC–O) has allowed the dif-
ferentiation of the odor-impact compounds from the background aroma-
active compounds in a fruit flavor. GC–O and GC–MS/GC–O rely on the
sniffing from the carrier gas stream coming out of the capillary column in
a GC system. Identification of the odor properties of a chemical and mass
spectra can be obtained based on the detector used (FID or MS).

Several methods can be used to quantify the intensity of a chemical compound. Among them, aroma extract dilution analysis (AEDA) is
a step-wise dilution of an extract with an organic solvent and is the most
employed in studies of fruit flavors. The intensity of significant odorants
is calculated by determining the flavor dilution factor (FD), which indicates the number of solvent dilutions of the aroma extract until the aroma
value is reduced to 1. Aroma extract concentration analysis (AECA) is the
opposite of AEDA and involves sequential concentration of an extract to
determine the most intense aroma-active compounds; however, it is not
commonly used due to column overload and changes in the sensorial
descriptors of some odorants at higher concentrations.

Regardless of the accuracy and sophistication of the technique used to
isolate and analyze the fruit of interest, several considerations need to be
taken into account. First, the isolation technique chosen needs to extract
the complete aroma profile of a fruit. The material needed for extraction
can come from the fruit itself—pulp, puree, oil, extract, absolute, essence,
juice, or concentrate. Besides the fruit itself, concentrates and purees are
the preferred substrates for analysis. However, concentration steps to
evaporate the solvent can create disturbance in the balance between the
volatile organic compounds that constitute the aroma of the fruit or even
provide oxidation or artifacts that are not found in the original crop of
interest.

Second, it is important to select an appropriate flavor isolation technique that produces the best yield of a compound without producing artifacts or breakdown intermediates that are not available in natural fruits.
Enzymes can significantly alter the concentration of these compounds and,
as a result, misrepresent the flavor profile of a tropical fruit. Degradation
or reaction capacity of sulfur compounds, significant contributors to
ripening note of several tropical fruits can significantly alter the overall
outcome of the flavor characterization of the fruits.

Finally, climacteric fruits require optimization of conditions for analysis because ripening conditions such as temperature and time will affect

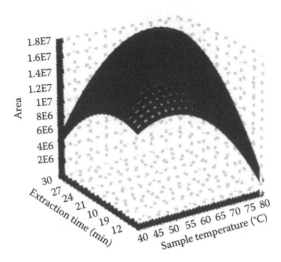

Figure 6.3 Response surface plot for total peak area for all compounds eluted versus sample temperature (°C) and extraction time (min). (From Casarek and Pawliszyn. 2006. *J Agric Food Chem.* 54:8688–8696.)

the composition of volatiles at each ripening stage (Liu and Yang 2002). Several research groups have studied the factors affecting the flavor isolation of chemical compounds in fruits by SPME and created response surface correlations that allow them to obtain the best yields for a specific fruit. Response surface method allows of the optimization of temperature and time during the analysis by SPME, thus controlling the effect of these variables on the vapor pressure and equilibrium of the compounds in the SPME vial. Figures 6.3 and 6.4 describe a factorial arrangement used for the optimization of the conditions to extract the volatile compounds of banana (Liu and Yang 2002) and other tropical fruits (Casarek and Pawliszyn 2006).

Main aroma compounds of some exotic tropical fruits

It is difficult to provide a range of flavor profiles that tropical fruits can cover because there are constantly new crops being added to this group with unique flavor profiles.

The most studied tropical fruits are mango and passion fruit, and their consumption in the U.S. market is high. The beverage market has a strong demand for these two fruits and other tropical flavors due to its current shift of sales from carbonated beverages to bottled water and functional categories due to concern with nutritional issues (Da Costa 2010). Even though there is more interest for tropical fruits with distinct

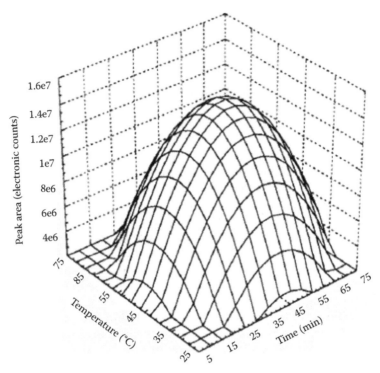

Figure 6.4 Response surface for the effect of time and temperature on the total peak area of banana flavor compounds. (From Liu T.-T., and Yang T.-S. 2002. *J Agric Food Chem* 50:653–657.)

flavor, scarce information is compiled about the flavor profile of tropical fruits with commercial interest beyond the analytical exercise of listing their compounds. The aim of this chapter is not to extend the list of chemicals identified in tropical fruits, but to provide a quick guidance about the flavor profile of tropical fruits that have a strong industrial potential and the aroma-active compounds that have been perceived, not only identified, as responsible for that flavor. Special consideration is given to using olfactometry data that considers the real contribution of the volatile to the aroma of a fruit, because the threshold difference between compounds makes quantified compounds to contribute differently to the characteristic aroma of a fruit. Furthermore, some compounds that might be perceived at the sniffing port could not have any corresponding peak at the MS because they might be present at levels below the detection level of the mass spectrometer.

The main aroma-active components of a selected group of tropical fruits with economic potential are summarized in Table 6.1.

Mango

Mango (*Mangifera indica* L.) is one of the most popular tropical fruits due to its different sensorial attributes. Extensive studies of the volatile constituents in mango have identified more than 300 free volatiles (Shibatomo and Tang 1990) and about 70 glycosidically bound compounds (Pino et al. 2005).

Steam distillation methods under normal and reduced pressure, static and dynamic headspace as well as SPME are the methods commonly used for analyzing this fruit. However, sample preparation has also been reported as causing significant variation in volatiles of this fruit. Aroma of mango varies widely among cultivars, and one typical composition is inaccurate. Common important volatile compounds contributors to the mango flavor reported by Pino et al. (2005), who studied the major volatile components of 20 mango cultivars, are terpenes (β-ocimene, d-3 carene), sesquiterpenes (β-caryophyllene, humulene, selinene), esters such as ethyl acetate and ethyl butanoate. These two esters, especially ethyl butanoate, have been clearly reported as the fruity note in Brazilian mangoes (Lopes, Fraga, and Rezende 1999). Straight–chain ester formation through oxidation of fatty acids has been reported in mango. Biosynthesis of fatty acids, the precursors of esters, has been found to be high during ripening of mango. Other esters commonly found in this fruit are ethyl 2-methylpropanoate and 2-methylbutanoate as well as ethyl (Z)-3-hexenoate and butyl 2-butenoate, the latter being with a characteristic green mango note. Important lactones are γ-octalactone (coconut-like odor) and carbonyls hexanal, (E)-2-hexenal, β-ionone as well as mesifuran (2,5-dimethyl-4-methoxy-3(2)H-furanone) and Furaneol (2,5-dimethyl-4-methoxy-3(2H)-furanone). One of the major discrepancies in the flavor of mango is related to sulfur notes, where diallyl disulfide, diallyl trisulfide, and 1,2-benzothiazole have been identified by distillation–extraction, while ethyl 3-mercaptobutyrate has been reported as odor-impact in this fruit (Dewis, Kendrick, and Swift 2002). Considerable differences in this group might be due to temperature degradation of sulfur-containing compounds or trace concentrations of these compounds under the detection limit of the detector employed in these studies.

Passion fruit

Passion fruit (*Passiflora edulis f. flavicarpa*) is one of the most popular tropical fruits in the food industry because of its characteristic fruity and sulfury flavor and strong aroma, which make it an excellent candidate for beverage applications. There are two main types of passion fruit: yellow and purple passion fruits. The yellow type is generally larger than the purple one, but the pulp of the purple type is less acidic and richer in aroma and

Table 6.1 Major Aroma-Active Contributors of the Flavor of Tropical Fruits with Economic Potential

Fruit	Botanic characteristics	Origin	Cultivation	Flavor	Odor-impact flavor compounds	Reference
Acerola	*Malpighia punicifolia L.*	Central America, Caribbean, Amazonian region	Caribbean Islands (Cuba, Barbados)	Cherry-like with some fruity and creamy notes	Furfural, hexadecanoic acid, 3-methyl-3-butenol, methyl hexanoate, methyl 3-methyl butanoate, ethyl hexanoate, limonene	Pino and Marbot (2001)
Araca-boi	*Eugenia stipitata*	Brazil	Brazil	Herbal with estery notes	Germacrene D and B, α-pinene, hexyl acetate, *cis*-3 hexenyl acetate	Franco and Shibamoto (2000)
Babaco	*Carica pentagona H.*	Ecuador	Ecuador, Israel, New Zealand, Middle Eastern countries	Overtones of berries, pineapple, and papaya	Ethyl (E)-2-but-2-enoate, butyl butanoate, ethyl hexanoate, ethyl butanoate, geraniol, β-damascenone, butan-1-ol, hexan-1-ol, octan-1-ol, diacetyl, benzyl isothiocyanate	Shaw, Allen, and Visser (1985); Bartley (1988); Barbeni, et al. (1990)
Banana	*Musa acuminata/ Musa balbisiana*	Southeast Asia	Guatemala, Costa Rica, Ecuador	Fruity creamy note with alcohol background	Isoamyl acetate, isobutyl butyrate, hexanal, isoamyl alcohol	Liu and Yang (2002)
Cherimoya	*Annona cherimola Mill.*	Ecuador, Colombia, Peru	Mediterranean, South America	Strawberry/ banana notes with brown sugar/ creamy notes	Methyl butanoate, butyl butanoate, 3-methylbutyl butanoate, 3-methylbutyl 3-methylbutanoate, and 5-hydroxymethyl-2-furfural	Ferreira, Perestrelo, and Câmara (2009)

(Continued)

Table 6.1 Major Aroma-Active Contributors of the Flavor of Tropical Fruits with Economic Potential (*Continued*)

Fruit	Botanic characteristics	Origin	Cultivation	Flavor	Odor-impact flavor compounds	Reference
Cupuacu	*Theobroma grandiflorum Spreng.*	Amazonian region	North Brazil, Bolivia, Peru	Estery flavor with creamy, fatty background	Ethyl 2-methyl butyrate, ethyl butyrate, butyl 2-methyl butyrate, ethyl hexanoate, acetoin and hexadecanoic acid	Franco and Shibamoto (2000)
Durian	*Durio sp. L.*	Thailand, Indonesia, Malaysia	Southeast Asia	Fruity with exotic sulfury notes	3,5-Dimethyl-1,2,4 trithiolane, ethyl 2-methyl butanoate, hexyl hexanoate, propyl-2-methyl butanoate, dimethyl sulfide	Weenen, Koolhaas, and Apriyantono (1996)
Guava	*Psidium guajava L.*	Central and South America	Northern South America	Jammy, sweet, fruity (blueberry-like) flavor quince-banana	Ethyl acetate, acetic acid, ethyl butyrate, isobutyl acetate, ethyl hexanoate, ethyl benzoate, ethyl octanoate, (Z)-3-hexenol, (Z)-3-hexenyl hexanoate, hexyl acetate, methyl benzoate, limonene, 3-penten-2-ol, 2-methyl-1-butanol, acetoin, 1-octanol, α-copaene, 1-octen-3-ol, 2-phenyl ethyl alcohol, pentane-2-thiol, 3-methylthio-hexyl butyrate	Yen and Lin (1999); Jordan et al. (2002); Bassols and Demole (1994)

Kiwifruit	*Actinidia deliciosa L.*	Southern China	New Zealand, South Africa, California, Chile	Fresh green to estery green	(E)-2-Hexenal, 3-penten-2-ol, hexanal, (E,E)-2,6-nanadienal, ethyl butanoate, 6-methyl-5 hepten-2-one, hexyl hexanoate, methyl benzoate, α-terpineol, methyl (2-methylthio)acetate	Young, Paterson, and Burns (1983); Jordan et al. (2002); Paterson, Macrae, and Young (1991); Rousell and Leahy (1995)
Langsat	*Lansium domesticum Corr.*	Southeast Asia	Thailand, Cambodia, Laos, Vietnam	Green-pear-like with herbal-citrus notes	(E)-2-hexenal, acetic acid, ethyl acetate, α-cubebene, limonene, phenylacetate, hexyl acetate, phenyl ethyl alcohol, methyl ethyl acetate, phenyl ethyl acetate, methyl 2-hydroxy-3-methylpentanoate and methyl 2-hydroxy-4-methylpentanoate	Wong et al. (1994); Laohakunjit and Kerdchoechuen (2007)
Lychee	*Litchi chinensis Sonn.*	Southern China, Vietnam, Indonesia, Philippines	Asia	Rose floral with citrus and nutty-like notes	(Z)-Rose oxide, 2-phenyl ethanol, geraniol, β-damascenone, isobutyl acetate, linalool, isobutyl acetate, guaiacol, 2-acetyl 2-thiazoline, Furaneol, isovaleric acid, dimethyl disulfide and dimethyl trisulfide	Ong and Acree (1998a); Mahattanatawee et al. (2007)

(Continued)

Table 6.1 Major Aroma-Active Contributors of the Flavor of Tropical Fruits with Economic Potential (*Continued*)

Fruit	Botanic characteristics	Origin	Cultivation	Flavor	Odor-impact flavor compounds	Reference
Mango	*Mangifera Indica L.*	India	India, Brazil, West Indies, Mexico	Floral, fruity, herbal with creamy and sulfury notes	Ethyl-2-methylpropanoate, ethyl acetate, ethyl butanoate, (E,Z)-2,6-nonadienal, butyl 2-butenoate, (E)-2-nonenal, methyl benzoate, β-ionone, β-ocimene, β-caryophyllene, decanal, 2,5-dimethyl-4-methoxy-3(2H)-furanone, Furaneol, γ-octalactone, diallyl disulfide, ethyl 3-mercaptobutyrate	Malundo et al. (1997); Lopes, Fraga, and Rezende (1999); Pino et al. (2005); Pino and Mesa (2006)
Mangosteen	*Garcinia mangostana L.*	Indonesia	Southeast Asia	Fruity green with acid taste	(E)-2-Hexenal, hexyl acetate, (Z)-3-hexenol, copaene, hexyl-n-valerate, 2-methyl-3-buten-2-ol, 2,2-dimethyl-4-octanal	MacLeod and Pieris (1982); Laohakunjit and Kerdchoechuen (2007)
Naranjilla	*Solanum quitoense Lam.*	Ecuador, Colombia and Central America	Andes and Central America	Sweet as a mix of banana, pineapple, and strawberry	Methyl butanoate, ethyl acetate, ethyl butanoate, butanoic acid, methyl benzoate, methyl (E)-2-butenoate, γ-hexalactone, linalool, acetic acid	Brunke, Mair, and Hammerschmidt (1989); Acosta, Perez, and Vaillant (2009)

	Scientific name	Region	Region	Aroma	Compounds	References
Noni	*Morinda citrifolia*	Hawaiian and Tahitian islands	Hawai	Foul smell and soapy taste when fully mature	Octanoic acid, hexanoic acid, ethyl octanoate, ethyl hexanoate, methyl octanoate, methyl hexanoate, 3-methyl-3-buten-1-ol, S-methyl thioacetate, dimethyl disulfide, methyl 3-methylthiopropanoate	Wei, Ho, and Huang (2011)
Papaya	*Carica papaya L.*	Central and South America	Hawai, Central America	Fruity-green with musty notes, cantaloupe-like	Linalool, linalool oxides, benzaldehyde, ethyl acetate, nonanal, methyl salicylate, β-caryophyllene, geranyl acetate	Flath Robert et al. (1990); Shibatomo and Tang (1990)
Passion fruit	*Passiflora edulis f. flavicarpa*	Paraguay, Brazil, Northern Argentina	India, Sri Lanka, Ecuador, Brazil	Floral, estery aroma with an exotic tropical sulfur note	Ethyl butanoate, 2-methylbutyl hexanoate, hexyl hexanoate, phenyl ethyl alcohol, hexanoic acid, furfural, octanol, isoamyl alcohol, β-ionone, Furaneol, methyl and ethyl 3-(methylthio) propanoic acid	Werkhoff (1998)
Pineapple	*Ananas comosus L.*	Brazil	Thailand, Philippines, Brazil	Sweet with fruity, musty notes	Furaneol, methyl 2-methylbutanoate, ethyl 2-methylbutanoate, ethyl acetate, ethyl hexanoate, ethyl butanoate, ethyl 2-methylpropanoate, methyl hexanoate, and methyl butanoate	Takeoka et al. (1989); Elss et al. (2005)

(Continued)

Table 6.1 Major Aroma-Active Contributors of the Flavor of Tropical Fruits with Economic Potential (*Continued*)

Fruit	Botanic characteristics	Origin	Cultivation	Flavor	Odor-impact flavor compounds	Reference
Rambutan	*Nephelium lappaceum L.*	Southeast Asia	Southeast Asia, Indonesia, Philippines, Costa Rica	Fruity-sweet with sweaty, fatty-green notes	β-Damascenone, ethyl 2-methyl butyrate, (E,E)-2,6 nonadienal, (E)-2-nonenal, nonanal, 3,5-dimethyl-1,2,4-trithiolane	Ong, Acree and Lavin (1998b)
Soursop	*Annona muricata*	Amazon and West India	West Indies, South America	Custard-apple-like	Methyl hexanoate, methyl hexen-2-enoate, Farnesene, methylbut-2-enoate	MacLeod and Pieris (1981); Iwaoka et al. (1993)
Star fruit/ carambola	*Averrhoa carambola*	Sri Lanka	Southeast Asia and Malaysia	Cross between an apple and a grape	(E)-2-Hexenal, methyl benzoate, ethyl acetate, (Z)-3-hexenol, (Z)-2-hexenol, *n*-hexanol and several minor alcohols, esters, and terpenes	Frohlich and Schreier (1989); MacLeod and Ames (1990); Casarek and Pawliszyn (2006)
Tamarind	*Tamarindus indica L.*	Sudan, tropical Africa	Belize, Central America, Brazil	Warm, citrus-like notes and some roasted undertones	2-Acetyl furan, 2-phenylacetaldehyde, 5-methyl-2-furfural, geraniol, geranyl acetate, nonanal, limonene, benzyl benzoate	Lee, Swords, and Hunter (1975); Casarek and Pawliszyn (2006)

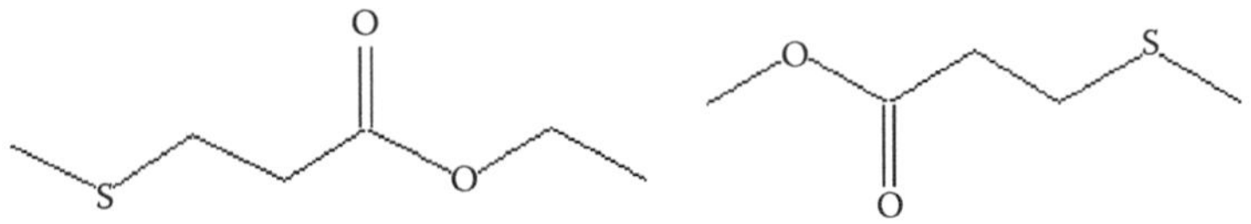

(a)   (b)

Figure 6.5 Structures of (a) ethyl 3-(methylthio)propionate and (b) methyl 3-(methylthio)propionate.

flavor, and has a higher proportion of juice. Because of its stronger flavor and aroma, the purple passion fruit is preferred for consumption as fresh juice, whereas the yellow passion fruit is better suited for processing.

Extensive studies have been conducted in both yellow and purple passion fruit as well as some of their hybrids (Casimir et al. 1981). Several techniques have been used to properly isolate the trace sulfur-containing components of passion fruit. For instance, Werkhoff et al. (1998) compared several headspace and distillation techniques under normal and reduced pressure to isolate the sulfur-containing compounds. In his research, vacuum solvent distillation extractions of the fruit showed more fruity and sulfury/exotic characteristics than atmospheric solvent extraction. Suppression of browning reactions under vacuum was believed to be the main reason for this discovery. Furthermore, the main compounds responsible for the exotic sulfury note of this fruit were the methyl, ethyl, butyl, pentyl, hexyl, and (Z)-3-hexenyl esters of 3-(methylthio)propanoic acid. Figure 6.5 illustrates the structures of the ethyl and methyl 3-(methylthio) propanoic acids.

Among the predominant compounds responsible for the fruity notes for yellow passion fruit were ethyl butanoate, 2-methylbutyl hexanoate, hexyl butanoate, hexyl hexanoate, ethyl 3-hydroxybutanoate, and 3-(Z)-hexenyl hexanoate, β-ionone, 2-phenylethyl butyrate, and 2,5-dimethyl-4-hydroxy-3(2H)-furanone (Furaneol).

Guava

Guava (*Psidium guajava*) is another tropical fruit with a wide range of applications currently in the industry. It is mainly consumed as puree, whole fruit, or processed juice products (Yen and Lin 1999). Most of the research has been conducted to study the flavor profile of two varieties white and pink guava. A study conducted by Jordan et al. (2002) identified the main components of guava puree by GC–O as ethyl acetate, acetic acid, ethyl butyrate, acetoin, isobutyl acetate, (Z)-3-hexenol, ethyl hexanoate, (Z)-3-hexenyl hexanoate, limonene, 2-phenyl ethyl alcohol. In addition, the study found that the differences between the flavor profiles of essence and puree of this fruit were based mainly on the stronger impact of E-2-hexenal on the essence rather than on the fruit puree.

Kiwi

Kiwifruit (*Actinidia deliciosa L.*) flavor can be described as a green fruity note, mostly due to the flavor impact of (E)-2-hexenal, with a fruity profile mainly caused by butyric acid esters. Fruity profile, butyric acid esters can dominate its profile. In addition to these compounds, α-pinene, β-pinene, linalyl acetate, isobornyl acetate, and α-terpenyl acetate are associated with overripening of this fruit (Jordan et al. 2002).

Several studies of the kiwifruit flavor have identified typical C6 compounds characteristic of the degradation of unsaturated fatty acids, but its concentration is altered due to an increase in terpene and butyric esters during ripening of the fruit. For instance, stored kiwifruit from later harvests had higher levels of α- and β-pinene than fresh fruit or immature stored fruit (Paterson, Macrae, and Young 1991).

There are more than 400 different varieties of kiwifruit solely in China where they have been used for over 700 years. However, differences in the sensory evaluations of kiwifruit varieties are partly due to the different proportions of the same volatile components and are also due to the presence of cultivar-specific impact compounds (Friel et al. 2007).

Lychee

Aroma of its flesh has been described as "rose and fruity-floral with cherry/citrus like notes" (Ong et al. 1998). However, a diversity of flavor profiles has been reported for this fruit due to the considerable number (>49) of cultivars for this fruit. The freshly picked lychee is strongly suggestive of sulfur compounds, but only a few sulfur compounds have been identified (dimethyl sulfide, dimethyl trisulfide, and 2-methyl thiazole) (Mahattanatawee et al. 2007).

The flavor of lychee can be characterized by the interaction between floral compounds (*cis*-rose oxide, 2-phenyl ethyl alcohol), citrus and fruity compounds (geraniol, linalool, isobutyl acetate), and some woody (guaiacol and acetyl-2-thiazoline) and sweet (Furaneol) notes.

Rambutan

Rambutan (*Nepthelium lappaceum. L*) belongs to the same family Sapindaceae as lychee and longan; however, this crop has been described as being less aromatic than its relatives. Its flavor is described as the interaction of fruity-sweet, fatty-green, and woody and spicy notes (Ong et al. 1998). The major contributors to the flavor of rambutan are β-damascenone (fruity-floral); ethyl 2-methyl butyrate (fruity-berry); Furaneol (strawberry furanone); some lipid oxidation compounds (E)-2-nonenal, nonanal, (E,E)-2,6-nonadienal (fatty-green); and some woody (4-,5-epoxy-(E)-2-decenal) and vanilla (vanillin) notes.

Pineapple

Pineapple (*Ananas comosus L.*) was originally discovered in Brazil; however, the large producers of this fruit are Thailand and Philippines with a combined 30% of the total cultivated area of this fruit. Several research groups have investigated the flavor of pineapple, which is mainly composed of methyl and ethyl esters (ethyl 2-methylbutanoate, methyl hexanoate), thioesters [ethyl 3-(methylthio)propionate and methyl 3-(methylthio)propionate] in combination with some lactones (gamma-octalactone and delta-octalactone) as well as the characteristic sugar/sweet notes provided by 2,5-dimethyl-4-methoxy-3(2H)-furanone (mesifurane) and 2,5-dimethyl-4-hydroxy-3(2H)-furanone (Furaneol) (Takeoka et al. 1989; Elss et al. 2005).

Papaya

Papaya or pawpaw (*Carica papaya L.*) is a crop seen for the first time in Central America. Its distinct flavor and diverse uses from medicine to meat tenderizer have made papaya one of the most common tropical fruits. However, papaya is a fruit considerably sensitive to the fruit fly (*Drosophila* sp.) and the Papaya ringspot virus. Most of the cultivars used in commercial plantations are genetically modified in order to protect against this virus, which has been found to have almost wiped out this fruit from the Hawaiian commercial plantations. The flavor of papaya is characterized by linalool, linalool oxide, benzaldehyde, ethyl acetate, nonanal, methyl salicylate, caryophyllene, geranyl acetate (Flath Robert et al. 1990).

Besides the commonly known tropical fruits, such as papaya, guava, kiwifruit, pineapple, mango, or passion fruit, other not-so-popular tropical fruits are being studied due to their health benefits and exotic flavor. A few examples of this constantly growing group are araca-boi, acerola, soursop, babaco, and naranjilla among others described in Table 6.1.

Araca-boi

Araca-boi (*Eugenia stipitata*) is a yellow skin round fruit with a creamy white pulp. Its flavor is a strong acidic green floral composed mainly of the sesquiterpenes (germacrene B and D) as well as some terpenes (α-terpinene and β-caryophyllene) and some C6 esters (*cis*-3-hexenyl acetate and hexyl acetate).

Acerola

Major constituents of acerola (*Malpighia punicifolia*) obtained by simultaneous steam distillation were furfural, hexadecanoic acid, 3-methyl-3-butenol, and limonene. Furfural, 3-hydroxy-2-pyranone, and some other compounds reported as volatile constituents of acerola were identified as degradation or oxidation of ascorbic acid during distillation (Pino and Marbot 2001).

Soursop

Soursop (*Annona muricata*) belongs to the Annonaceae group. It has a less sweet flavor than other members of its group. Its flavor is mainly characterized by saturated and unsaturated methyl and ethyl esters of six and eight carbons as well as β-farnesene. Among them, methyl hexanoate and methyl hex-2-enoate are the more intense compounds of the soursop flavor (MacLeod and Pieris 1981).

Babaco

Babaco (*Carica pentagona H.*) is originally from the warm microclimates at high altitudes in Ecuador. This tree belongs to the Caricaceae family and, therefore, is a relative to papaya (*Carica papaya*). Nevertheless, it has a mild and aromatic fruity estery flavor profile different from papaya. The main contributors to the babaco flavor are ethyl (E)-2-but-2-enoate, butyl butanoate, ethyl butanoate, ethyl hexanoate, geraniol, β-damascenone, butan-1-ol, hexan-1-ol, octan-1-ol, diacetyl, and benzyl isothiocyanate.

Naranjilla

The flavor of naranjilla (*Solanum quitoense Lam.*) has been described as sweet, similar to that of a mixture of banana, pineapple, and strawberry (Brunke, Mair, and Hammerschmidt 1989). Naranjilla, also known as lulo, is a fruit commonly used in northern countries of South America, mainly as concentrates, purees, and juices. Commercial plantations are found in Colombia, Ecuador, and Peru. The main compounds responsible for the flavor of naranjilla are methyl butyrate, ethyl acetate, methyl (E)-2-butyrate, linalool, γ-hexalactone, α-terpineol, and acetic acid (Silva, Suarez, and Duque 1990).

Summary

This chapter has aimed to cover the isolation and characterization of the main aroma compounds responsible for the flavor of tropical fruits with economic potential. These fruits are commonly inexpensive and extremely rich in vitamins and can be used in a variety of products. Due to the climatic requirements for this group of fruits, more crops known only regionally are being studied and incorporated to this diverse group. Although the aroma-characterizing chemicals for some tropical fruits of commercial interest have been described in this chapter, by no means the variation within varieties of these crops has been covered. Special consideration was also given to identification of the major contributors to the flavor of these fruits using GC–O data instead of a list of tentatively identified compounds from an extract.

Particularly for fruits and fruit concentrates or juices the techniques commonly used to trap their flavor profiles are vacuum steam distillation, liquid–liquid extraction, and solvent-assisted flavor extraction. Headspace and sorptive techniques (HSSE and SBSE) are preferred over distillation and extraction techniques to avoid temperature-degradation artifacts as well as to reduce time required for analysis. Headspace analyses are mainly performed by static or dynamic headspace (purge and trap) as well as by SPME. Combination of olfactometry and GC allows the selection of the major contributors to the characteristic flavor of tropical fruits from a chemical identification list.

In general, the distinct flavor of tropical fruits is mostly due to the balance between ethyl and methyl esters, terpenes, sesquiterpenes, aldehydes, alcohols, and sulfur compounds. However, the ratio and characteristic aroma profile of climacteric tropical fruits are affected by ripening stage and cultivation conditions. Potential understanding of the biosynthesis of these compounds as well as the formation of off-flavors is important for the industry to create tailored flavor profiles with high commercial values as well as preservation methods for these characterizing aroma-active chemicals.

References

Acosta, O., A. M. Perez, and F. Vaillant. 2009. Chemical characterization, antioxidant properties and volatile constituents of naranjilla (Solanum quitoense Lam.) cultivated in Costa Rica. *Arch Latinoam Nutr* 59(1):88–94.

Barbeni, M., P. A. Guarda, M. Villa, P. Cabella, F. Pivetti, and F. Ciaccio. 1990. Identification and sensory analysis of volatile constituents of babaco fruit (Carica pentagona heilborn). *Flavour Frag J* 5(1):27.

Bartley, J. P. 1988. Volatile flavor components in the headspace of the babaco fruit (Carica Pentagonia). *J Food Sci* 53(1):138–140.

Bassols F., and E. P. Demole. 1994. The occurrence of pentane-2-thiol in guava fruit. *J Essent Oil Res* 6(5):481–483.

Brunke, E. J., P. Mair, and J. Hammerschmidt. 1989. Volatile from naranjilla fruit (Solanum quitoense Lam.). GC–MS analysis and sensory evaluation using sniffing GC. *J Agric Food Chem* 37(3):746–748.

Casarek, E., and J. Pawliszyn. 2006. Screening of tropical volatile compounds using Solid-Phase Microextraction (SPME) fibers and internally cooled SPME Fiber. *J Agric Food Chem* 54:8688–8696.

Casimir, D. J., J. F. Kefford, and F. B. Whitfield. 1981. Technology and flavor chemistry of passion fruit juices and concentrates. *Advances in Food Research* 27:243–245.

Da Costa, N. C., ed. 2010. *Overview of Flavors in Noncarbonated Beverages Flavors in Noncarbonated Beverages*. Washington, D.C.: American Chemical Society.

Da Costa, N. C., and E. Sanja. 2005. Identification of Aroma Chemicals. In *Flavors and Fragrances*. ed. D. J. Rowe Poole, 12–31. UK: Blackwell.

Dewis, M. L., L. Kendrick, and K. A. D. Swift. 2002. Creation of flavours and the synthesis of raw materials inspired by nature. In *Advances in Flavours and Fragrances: From the Sensation to the Synthesis*, ed. K. A. D. Swift, 147. London: The Royal Society of Chemistry.

Elss, S., C. Preston, C. Hertzig, F. Heckel, E. Richling, and P. Schreier. 2005. Aroma profiles of pineapple fruit (Ananas comosus [L.] Merr.) and pineapple products. *LWT—Food Sci Technol* 38(3):263.

Fernandes, A. F., S. Rodrigues, L. L. Chung, and S. A. Mujumdar. 2011. Drying of exotic tropical fruits: A comprehensive review. *Food Bioprocess Technol* 4:163–185.

Ferreira, L., R. Perestrelo, and J. S. Câmara. 2009. Comparative analysis of the volatile fraction from Annona cherimola Mill. Cultivars by solid-phase microextraction and gas chromatography-quadrupole mass spectrometry detection. *Talanta* 77(3):1087.

Flath Robert, A., D. M. Light, E. B. Jang, and J. O. John. 1990. Headspace examination of volatile emissions from Ripening Papaya (Carica papaya L. solo variety). *J Agric Food Chem* 38(4):1061–1063.

Fleuriet, A., and J. J. Macheix. 2003. Phenolic acids in fruits and vegetables. In *Flavonoids in Health and Disease*, ed. R.-E. L. Packer, 1–36. New York: Marcel Dekker.

Franco, M. R. B., and T. Shibamoto. 2000. Volatile composition of some Brazilian fruits: Umbu-caja (Spondias citherea), Camu-camu (Myrciaria dubia), Aracaboi (Eugenia stipitata), and Cupuacu (Theobroma grandiflorum). *J Agric Food Chem* 48(4):1263.

Friel, E. N., M. Wang, A. J. Taylor, and E. A. MacRae. 2007. In vitro and in vivo release of aroma compounds from yellow-fleshed Kiwifruit. *J Agric Food Chem* 55(16):6664.

Frohlich, O., and P. Schreier. 1989. Additional volatile constituents of carambola (Averrhoa carambola L.) fruit. *Flavour Frag J* 4(4):177.

Halliwell, B., M. A. Murcia, S. Chirico, and O. I. Aruoma. 1995. Free radicals and antioxidants in food and in vivo: what they do and how they work. *Crit Rev Food Sci Nutr* 35(1-2):(7–20).

Iwaoka, W. T., Z. Xiaorong, R. A. Hamilton, C. L. Chia, and C. S. Tang. 1993. *Identifying Volatiles in Soursop and Comparing their Changing Profiles during Ripening*. Alexandria, VA: ETATS-UNIS: American Society for Horticultural Science.

Jordan, M., A. C. Margaria, E. P. Shaw, and K. L. Goodner. 2002. Aroma active components in aqueous Kiwi fruit essence and Kiwi fruit puree by GC–MS and multidimensional GC/GCO. *J Agric Food Chem* (50):5386–5390.

Laohakunjit, N., and O. Kerdchoechuen. 2007. Postharvest survey of volatile compounds in five tropical fruits using headspace-solid phase microextraction (HS-SPME). *HortSci.* 42(2):309–314.

Lee, P. L., G. Swords, and G. L. K. Hunter. 1975. Volatile constituents of tamarind (Tamarindus indica). *J Agric Food Chem* 23(6):1195.

Liu, T.-T., and T.-S. Yang. 2002. Optimization of solid-phase microextraction analysis for studying of headspace flavor compounds of banana during ripening. *J Agric Food Chem* 50:653–657.

Lopes, D. C., S. R. Fraga, and C. M. Rezende. 1999. Aroma impact substances on commercial Brazilian mangoes by HRGC-O-AEDA-MS. *Quim. Nova* 22:31–36.

MacLeod, G., and J. M. Ames. 1990. Volatile components of starfruit. *Phytochem* 29(1):165.

MacLeod, A. J., and N. M. Pieris. 1981. Volatile flavor components of soursop (Annona muricata). *J Agric Food Chem* 29:488–490.

MacLeod, A. J., and N. M. Pieris. 1982. Volatile flavour components of mangosteen, Garcinia mangostana. *Phytochem* 21(1):117.

Mahattanatawee, K., P. Ruiz Perez-Cacho, T. Davenport, and R. Rouseff. 2007. Comparison of three lychee cultivar odor profiles using gas chromatography-olfactometry and gas chromatography-sulfur detection. *J Agric Food Chem* (55):1939–1944.

Malundo, T. M. M., E. A. Baldwin, M. G. Moshonas, R. A. Baker, and R. L. Shewfelt. 1997. Method for the rapid headspace analysis of Mango (Mangifera indica L.) homogenate volatile constituents and factors affecting quantitative results. *J Agric Food Chem* 45(6):2187.

Ong, P. K. C., and T. E. Acree. 1998a. Gas chromatography/olfactory analysis of Lychee (Litchi chinesis Sonn.). *J Agric Food Chem* 46(6):2282.

Ong, P. K. C., T. E. Acree, and E. H. Lavin. 1998. Characterization of volatiles in Rambutan fruit (Nephelium lappaceum L.). *J Agric Food Chem* (46):611–615.

Paterson, V. J., E. A. Macrae, and H. Young. 1991. Relationships between sensory properties and chemical composition of kiwifruit (Actinidia deliciosa). *J Sci Food Agric* 57(2):235.

Pino, J. A., and R. Marbot. 2001. Volatile flavor constituents of acerola (Malpighia emarginata DC.). *J Agric Food Chem* (49):5880–5882.

Pino, J. A., and J. Mesa. 2006. Contribution of volatile compounds to mango (Mangifera indica L.) aroma. *Flavour Frag J* 21(2):207.

Pino, J. A., J. Mesa, Y. Munoz, M. P. Marti, and R. Marbot. 2005. Volatile components from Mango (Mangifera indica L.) cultivars. *J Agric Food Chem* 53(6):2213.

Reineccius, G. 2005. Flavor formation in Fruits and Vegetables. In *Flavor Chemistry and Technology*, ed. G. Reineccius, 76–103. New York: Taylor & Francis.

Rufino, M. S., A. F. Fernandes, R. E. Alves, and E. S. Brito. 2009. Free-radical-scavenging behaviour of some north east Brazilian fruit in a DPPH system. *Food Chem* (114):693–695.

Shaw, G. J., J. M. Allen, and F. R. Visser. 1985. Volatile flavor components of babaco fruit (Carica pentagona, Heilborn). *J Agric Food Chem* 33(5):795.

Shibatomo, T., and S. Tang. 1990. 'Minor' tropical fruits-mango, papaya, passion fruit, and guava. In *Food Flavours: Part C. Amsterdam*, eds. A. Morton and A. J. MacLeod, 221–234. The Netherlands: Elsevier.

Silva J., M. Suarez, and C. Duque. 1990. Preparation of a lulo (Solannun vestissimun D.) essence from the main volatile components of the fruit. *Rev Colomb Quim* 19(2):47–54.

Takeoka G., G. R. Buttery, A. F. Flath, R. Teranishi, E. L. Wheeler, R. L. Wieczorek, and M. Guentert. 1989. Volatile Constituents of Pineapple *Ananas Comosus [L.] Merr.*). In *Flavor Chemistry*, ed. R. Teranishi, R. G. Buttery and F. Shahidi, 223–237. American Chemical Society.

Weenen, H., W. E. Koolhaas, and A. Apriyantono. 1996. Sulfur-containing volatiles of Durian Fruits (Durio zibethinus Murr.). *J Agric Food Chem* 44:3291–3203.

Flavor, fragrance, and odor analysis

Wei, G.-J., C.-T. Ho, and A. S. Huang. 2011. Analysis of volatile compounds in Noni Fruit (Morinda citrifolia L.) juice by steam distillation-extraction and solid phase microextraction coupled with GC/AED and GC/MS. *J Food & Drug Anal* 19(1):33–39.

Werhoff, P., M. Guentert, G. Krammer, H. Sommer, and J. Kaulen. 1998. Vacuum headspace method in aroma research: Flavor chemistry of yellow Passion Fruits. *J Agric Food Chem* 46:1076.

Wong, K. C., S. W. Wong, S. S. Siew, and D. Y. Tie. 1994. Volatile constituents of the fruits of lansium domesticum correa (Duku and Langsat) and baccaurea motleyana (Muell. Arg.) Muell. Arg. (Rambai). *Flavour Frag J* 9(6):319.

Yen, G.-C., and H.-T. Lin. 1999. Changes in volatile flavor components of guava juice with high-pressure treatment and heat processing and during storage. *J Agric Food Chem* 47(5):2082.

Young, H., V. J. Paterson, and D. J. W. Burns. 1983. Volatile aroma constituents of kiwifruit. *J Sci Food Agric* 34(1):81.

Rousell, R. L., and M. Leahy, eds. 1995. *Volatile Compounds Affecting Kiwifruit Flavors Fruit Flavors*. Washington, D.C.: American Chemical Society.

chapter seven

On the synthesis and characteristics of aqueous formulations rich in pyrazines

William M. Coleman, III

Contents

Introduction

The chemical reactions by which a large majority of the enjoyable food flavors and aromas are produced are termed "nonenzymatic browning" (Teranishi, Flath, and Sugisawa 1981). From a general perspective, this reaction is one that occurs between the endogenous carbohydrates and amino acids and/or ammonium ions in natural products. Among the products generated from the reaction are volatile materials of relatively low molecular weights that are responsible in part for the powerful aromas associated with heat-treated/cooked natural products (Clarke 1986; Shibamoto and Bernhard 1977; Teranishi, Flath, and Sugisawa 1981; Waller and Feather 1983; Buckholz 1988; Whitfield 1992; Coleman 1996; Leahy 1985). The qualitative and semiquantitative characteristics of headspace volatiles above a diverse array of heat-treated materials containing nitrogen sources such as amino acids and sugars have been described in the literature (Coleman and Perfetti 1997; Coleman 1996; Coleman and Steichen 2006). In addition, proposed reaction mechanisms underlying the synthesis of the volatile and semivolatile compounds have been established (Coleman 1999a,b; Shibamoto and Bernhard 1977; Chen and Ho 1998, 1999; Lee 1995; Parliament, Morello, and McGorrin 1994; Coleman and Steichen 2006). The low-molecular-weight volatile and semivolatile portions of heat-treated materials constitute only around 1% by weight of the final reaction mixture; yet with odor thresholds in the milligram–kilogram range, these compounds have significant impact on the acceptability of the mixture.

Three major categories of low-molecular-weight volatile and semivolatile compounds have been described in heated formulations containing sugars and nitrogen sources: (1) pyrazines, (2) Strecker aldehydes, and (3) products of thermal degradation of sugar. In addition, relatively minor compounds such as imidazoles, thiazoles, and oxazoles can also be present. Assessments of the amounts and types of these volatile compounds have been very useful in determining the course and extent of sugar/nitrogen reactions in a wide array of formulations. The powerful aroma associated with these compounds and similar heat-treated formulations has been attributed in part to the presence of pyrazines containing linear and branched alkyl substituents. Pyrazines containing branched alkyl substituents may be represented by (3-methylbutyl)pyrazine. The presence of pyrazines containing specific branched alkyl substituents has been unambiguously attributed to the presence of certain amino acids in the formulations (Hwang, Hartman, and Ho 1995; Chen and Ho 1998; Coleman 1999a,b). Structure–activity relationship studies have shown that the intensity and characteristics of the aroma of alkylpyrazines can be related to pyrazine molecular shape and topology. In general, as the alkyl substituents become longer and more branched, the odor intensity increases dramatically (Tsantili-Kakoulidou and Kier 1992).

To date, the vast majority of model reactions and natural products fortified with flavors that result in pyrazine-rich formulations upon heating have used sugars such as fructose, glucose, fructose–glucose mixtures, and rhamnose as components of the formulations. These sugars have been shown to serve as carbon sources for formation of the pyrazine aromatic ring structure (Shibamoto and Bernhard 1977; Coleman 1999a, b; Hwang, Hartman, and Ho 1995).

However, there is very little information available in the literature on the use of alternative molecules as pyrazine-ring building blocks. Several U.S. patents (Onishi, Nishi, and Kakizawa 1969; Suwa, Satoh, and Shida 1973; Warfield, Galloway, and Kallianos 1975; Wu and Swain 1981, 1983; Cox, Grubbs, and Haut 1987) disclose the use of hydroxyketone-like materials with amino acids, ammonium hydroxide, and aldehydes to produce "aromatic" tobaccos. These patents neither disclose the structures of the components that yield aromatic tobaccos nor describe the mechanisms underlying the production of such aromatic materials. It can be reasonably postulated that novel aromatic tobaccos contain enhanced levels of pyrazines. This suggestion is consistent with findings in the literature on proposed pyrazine reaction intermediates. For example, reaction pathways/ mechanisms have been proposed with hydroxyketones and aldehydes as intermediates in the synthesis of pyrazines (Shibamoto and Bernhard 1977; Buttery, Stern, and Ling 1994; Pons et al. 1991). Investigations were undertaken to examine the possibilities of pyrazine production associated with selected combinations of hydroxyketones, aldehydes, amino acids, and ammonium hydroxide (Coleman and Lawson 2000). The report describes the findings for a series of formulations containing hydroxyketones, aldehydes, amino acids, and ammonium hydroxide prepared in a microwave oven under a variety of process conditions, reagent types, and reagent ratios. The results of changes in these variables are discussed in terms of pyrazine yield and distribution.

This chapter describes in detail a number of the parameters utilized during the production of aqueous formulations rich in pyrazines. The parameters include, for example, concentration, reagent ratios, reaction time, reaction temperature, and reaction pressure. Also included in this set is a detailed description of the analytical procedures employed in the qualitative, quantitative, and relative quantitative analyses of the aqueous formulations.

Materials and methods

The following section will provide details of the analytical approaches employed during these studies. Information on, for example, gas chromatographic conditions, sample preparation and data analyses will be provided.

Headspace analysis

Dynamic headspace/gas chromatography/mass selective detection/flame ionization detection (GC/MSD/FID) was employed for headspace analyses. This approach was very similar to another that was reported earlier (Coleman, White, and Perfetti 1994, 1996; Coleman and Lawson 2000). More specifically, for each aqueous sample 1.0 mL was placed in a 5-mL sparge tube along with 1 mL of an aqueous standard containing 21.8 mg/L of cyclohexanone in deionized water as an internal standard. The yield of volatiles was calculated based on the response of the added cyclohexanone. Five measurements were usually made on each sample and the relative standard deviations were in the range ± 6%. The headspace sampling parameters generally used in the studies are listed in Table 7.1.

Table 7.1 Headspace–GC–MSD Operating Conditions

GC conditions	
System configuration	Hewlett Packard (HP) 5880 GC equipped with a 5970 MSD and a Tekmar LSC 2000 Autosampler
Column	DB-1701, 30 m, 0.32-mm i.d. (inside diameter), 1-µm film thickness
Injection port temperature	250°C
Injection	Splitless
Inlet pressure	20 psi
Column oven initial temperature	10°C
Column oven initial time	0 min
Column oven initial ramp 1 rate	2.5°C/min
Column oven 1 final temperature	47°C
Column oven ramp 2 rate	10°C/min
Column oven 2 final temperature	230°C
Column oven 2 final time	20 min
Headspace conditions	
Sample purge time	20 min
Sample preheat time	5 min
Sample desorb time	5 min
Sample desorb temperature	180°C
Sample purge temperature	70°C
Mass spectrometer transfer-line temperature	250°C
Mass spectral databases	NBS, Wiley
Mass spectrometer configuration	Electron impact, 70 eV

Source: Coleman III, W. M., and T. J. Steichen. 2006. *J Sci Food Agric* 86:380–391. Copyright Wiley-VCH Verlag GmbH & Co. KGaA. Reproduced with permission.

Automated solid-phase microextraction analysis

The aqueous samples were analyzed using automated solid-phase microextraction (AUTOSPME). A Varian (Pao Alto, California) 8200 CX Auto-Sampler with SPME III sample agitation was mounted atop a Hewlett Packard (HP; Pao Alto, California) 5890 Series II Plus GC equipped with an HP 5972 MSD operating either in the scan mode at 70 eV or in the selected ion monitoring (SIM) mode. The GC was fitted with a DB-Wax fused-silica column (30 m × 0.25 mm i.d. (inside diameter), 0.5-µm film thickness; J&W Scientific, Folsom, California). The MSD interface and GC injection port temperatures were 250°C. The GC oven was temperature programmed from 40°C to 140°C at 5°C/min, and then to 220°C at 10°C/min and held there for 4 min. The injection port was fitted with a narrow bore liner. Splitless injections were made and the split was opened after 1 min. The fiber was automatically submerged in the aqueous solution, vibrated for 0.75 min, removed, injected, and held in the injection port for 30 min. This 30-min holding time was selected to simplify the timing commands of various components of the instrument configuration. Fresh samples were used for every injection. Under these operating conditions, no fiber performance degradation was noted for at least 100 injections. After approximately 100 injections, the injection port liner was replaced. For six replicate injections of the same solution contained in six separate vials, the SIM area count for methylpyrazine had a percentage RSD value of 6.5.

The SPME fibers for these automated injections were obtained from Supelco (Bellefonte, Pennsylvania), and they were employed strictly following the manufacturer's instructions. Carbowax/divinylbenzene (CWDVB), carboxen/polydimethylsiloxane (carboxen/PDMS), PDMS/divinylbenzene (PDMS/DVB), and PDMS SPME fibers were evaluated much in the same manner as described in earlier publications (Coleman 1999a,b; Coleman and Lawson 1999). The PDMS SPME fiber having a film thickness of 100 µm was selected based on consistency of response, lack of measurable carryover, and extraction capability.

Prior to analysis by AUTOSPME/GC/MSD, any aqueous heat-treated suspension or was manually filtered through a Whatman (Florham, New Jersey) Autovial equipped with a 0.45-µm polyvinylidenefluoride (PVDF) filter designed for use with aqueous solutions. Then, to 1.8-mL vials equipped with Teflon-lined septa was added 1.7 mL of the filtered solution via a Rainin (Woburn, Massachusetts) EDP-Plus motorized microliter pipette. Strict attention to consistency in the addition of 1.7-mL filtered solution was necessary to obtain reproducible results. The volume 1.7 mL was selected to minimize the headspace inside a 1.8-mL vial and yet allow enough room for extensive movement of the solution during vibration of the fiber. No further experiments to optimize the vial volume were attempted. The charged vials were loaded on the sample carousel and automatically sampled employing the instrumentation software provided by Varian and HP. In some cases, it was necessary to dilute the heat-treated suspensions with water to obtain reproducible fiber performance.

Quantitative gas chromatography/selected ion monitoring–mass selective detection analyses

In some cases, SIM was used for the analysis of selected pyrazines and furans in the heat-treated formulations. The C2, C3, and C4 notations that appear before pyrazines were employed to denote classes of pyrazines. For example, C2 pyrazines include substituted pyrazines such as all the dimethylpyrazines as well as ethylpyrazine. In all these cases, the pyrazines have two carbons (C2) attached in some fashion to the fundamental pyrazine molecule. The C3 and C4 notations indicate attachment of three and four carbons, respectively. For specific quantitative analyses, tables of selected ions (m/z) indicative of a known pyrazine were constructed. A response factor for each known pyrazine was calculated from linear calibration curves having R^2 values > 0.999. The analysis conditions described for the SPME approach were employed for the quantitative GC/SIM–MSD approach. The sole exception was that 1-μL injections of a solution (split 10:1 and diluted with methanol 10:1) were performed instead of desorption of the exposed SPME fiber.

Sample preparation

This section will provide detailed information on the processes used to generate the aqueous formulations. In some cases, sealed vessels under pressure at selected temperatures were employed as reaction hardware and in other cases open vessels at ambient pressure and selected temperatures were used.

Microwave-scale quantities, sealed vessels

Heat treatment of aqueous samples or suspensions was conducted employing the microwave conditions described in Table 7.2. A typical aqueous

Table 7.2 Microwave Oven Operating Parameters

System	CEM Model MES-1000
Sample temperatures	Selected temperature in degree Celsius
Sample temperature ramp time	10 min
Microwave power	950 ± 50 W
Microwave frequency	2450 MHz
Sample heating times	Selected time in min
Ramp time to heating temperature	10 min

Source: Coleman III, W. M., and T. J. Steichen. 2006. *J Sci Food and Agric* 86:380–391. Wiley-VCH Verlag GmbH & Co. KGaA.

formulation was prepared as follows: To a microwave-permeable sealed reaction vessel 0.028 moles of 3-hydroxy-2-propanone, 0.028 moles of NH_4OH (30% aqueous ammonium hydroxide), and a selected amount of amino acid ranging from 0.5 to 2.0 g or a selected aldehyde ranging from 25 to 200 μL were added. The final volume of the reaction mixture was adjusted to 15 mL with water. For example, when examining the influence of the presence of a mixed fructose–glucose formulation, such as high-fructose corn syrup (HFCS) versus hydroxyketones in the formulation, the amount of water was adjusted to maintain a final reaction mixture volume of 15 mL. By doing so, the reactants/reactant mole ratios were maintained constant throughout the entire group of experiments.

Pilot-scale quantities, sealed vessels

Larger quantities of heat-treated aqueous formulations (≥ 1 L) were often prepared in sealed, stirred stainless steel reactors (Parr Instruments, Moline, Illinois). The conditions established as a result of the microwave reactions were translated into volumes and conditions for synthesis in sealed high-pressure reactors having volumes of 1 L, 2 gal, and 50 gal. The yields of pyrazines from these reactors were measured in the same manner as described in the section, *Quantitative gas chromatography/selected ion monitoring-mass selective detection.*

Open vessels

Using reagent ratios described in the section, *Microwave-scale quantities, sealed vessels,* reactions were performed in condenser-adapted open vessels using the boiling point of water as temperature control. Reaction times were controlled and selected volumes of aqueous condensate were collected and analyzed.

Reagents

The vast majority of reagents mentioned in this chapter were obtained from Aldrich Chemical Company (St. Louis, Missouri) and used as received. Deionized water (> 17.5 megohm/cm) was used. The HFCS was obtained from Corn Products International, Inc. (Bedford Park, Illinois), and used as received. The ^{15}N and ^{13}C reagents were obtained from CIL (Andover, Massachusetts). The majority of the pyrazines used were obtained from Pyrazine Specialties (Atlanta, Georgia).

Results and discussion

In this section, information on the qualitative and quantitative nature of the volatile and semi-volatile components contained within a variety of reaction configurations will be discussed. The influence of performing the reactions in sealed and open vessels will be addressed.

Qualitative and relative quantitative analyses, reactions in sealed vessels

A preliminary set of experiments to assess the impact of reaction variables and reagent ratios with respect to their impact on the formation of pyrazines in laboratory-scale aqueous formulations was conducted employing microwave process parameters. The results clearly indicated that microwave treatment of selected formulations produced aqueous compositions having enhanced levels of pyrazines. Examples of pyrazines formed are listed in Table 7.3. Selection of reagents in a formulation

Table 7.3 A Representative List of the Major Volatile Pyrazines Produced during Microwave Processing of Aqueous Formulations

Pyrazine
Methyl
2,5-Dimethyl
2,6-Dimethyl
Ethyl
2,3-Dimethyl
2-Ethyl-6-methyl
2-Ethyl-5-methyl
Trimethyl
Vinyl
Propyl
2-Methyl-5-(1-methylethyl)
2-Methylpropyl
2,6-diethyl
3-Ethyl-2,5-dimethyl
3-Ethyl-3,5-dimethyl
2,5-Diethyl
2,3-Diethyl
2,3-Dimethyl-5-ethyl
2-Ethyl-3,5-dimethyl
2-Vinyl-6-methyl
2-Vinyl-5-methyl
2,3,5-Trimethyl-6-ethyl
2,6-Dimethyl-3-vinyl (tentative)
2-Methyl-6-propyl

Table 7.3 A Representative List of the Major Volatile Pyrazines
Produced during Microwave Processing of Aqueous
Formulations (*Continued*)

2-Methyl-3-propyl
2-Methyl-3-(2-methylpropyl) isomer
Tetramethyl
3,5-Diethyl-2-methyl
2,3-Diethyl-6-methyl
2,5-Dimethyl-2-(2-methylpropyl) isomer
3,5,6-Trimethyl-2-(2-methylpropyl) isomer
2,3-Dimethyl-3-propyl
2,3-Dimethyl-5-propyl
2-Methyl-3-(1-propenyl) tentative
2-(2-Methylbutyl)
2-(3-Methylbutyl)
2-Methyl-3-(2-methylbutyl) isomer
2-Methyl-3-(3-methylbutyl) isomer
2-Methyl-6-(1-propenyl) tentative
2,5-Dimethyl-3-(2-methylbutyl) isomers
2,5-Dimethyl-3-(3-methylbutyl) isomers
2,5-Dimethyl-6,7-dihydro-5H-cyclopenta
Methylcyclopenta
Dimethylcyclopenta
2,3,5-Trimethyl-6-(2-methylbutyl)
2,5,6-Trimethyl-3-(2-methylpropyl)cyclopenta isomers
2,5,6-Trimethyl-3-(2-methylbutyl) isomers
2,5,6-Trimethyl-3-(3-methylbutyl) isomers

comprising HFCS (sugar), in addition to diammonium phosphate (DAP) or ammonium hydroxide (base), and selected amino acids afforded the opportunity to produce solutions having powerful, pleasant aromas and a diverse array of pyrazines. A typical formulation for these preliminary surveys would include 15 g of HFCS (sugar), 4 g of 30% NH_4OH, 11 g of water, and an amount of amino acid to yield an amino acid/sugar ratio of 0.37:1.00. Preliminary investigations of processing variables such as time, temperature, pH, and concentration likewise afforded solutions with selected distributions of pyrazines, amino sugars, and sugar thermal degradation products. Selection of amino acids played a critical role in controlling the amount and type of pyrazines formed. Such control affords the capability of preparing formulations with certain sensory characteristics. For example, formulations rich in dimethylethylpyrazines most likely have the most powerful impact due to the very low sensory threshold of these pyrazines relative to other pyrazines. Of particular note was the observation of the presence of pyrazines containing branched alkyl moieties in certain cases, particularly for the formulations containing leucine (LEU) and valine (VAL). Thus, the relationship between amino acid employed and pyrazine structure was established (Structure 7.1). The effect of reaction temperature on the yield of pyrazines (Figure 7.1) in these model aqueous formulations was found to be substantial. Note that as the processing temperature increased, the yield of pyrazines increased in all except one case (no nitrogen source). The amount of amino acid, expressed as a sugar/amino acid mole ratio, also influenced the yield of pyrazines (Figure 7.2). Further, the amino acid structure influenced the distribution of pyrazines, *vide supra, Relationship between amino acids and pyrazines found in microwave heat treated formulations.*

During heat treatment of the formulations, a decrease in pH was observed. From an initial value of approximately 7, pH values as low as 4.5 were observed for heat-treated formulations. This observation was of interest due to findings in the literature that indicated decreases in pyrazine yield in model systems with decreases in the pH levels of mixtures.

Thus, a series of microwave heat-treated formulations was prepared using concentrated buffer solutions. When employing a concentrated buffer designed to operate around a pH of 7, the pH of the casing formulation prior to heating was 6.8 and the pH after heating was found to be 6.3. Thus, reasonable control of pH was achieved using the buffer. Figure 7.3 shows the influence of the pH = 7 buffer on the yield of total pyrazines for serine (SER), threonine (THR), and SER–THR heat-treated formulations. The buffered reaction matrix resulted in a significant increase in the yield of total pyrazines in the SER, THR, and SER–THR formulations. In the

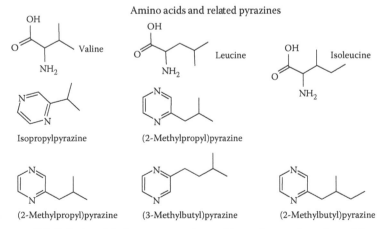

Structure 7.1 Relationship between amino acids and pyrazines found in microwave heat-treated formulations.

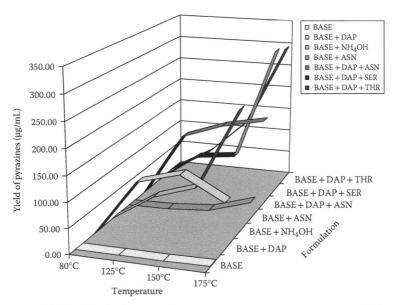

Figure 7.1 Yield of volatile pyrazines versus process temperature and formulation. ASN = asparagine, DAP = diammonium phosphate, SER = serine, and THR = threonine.

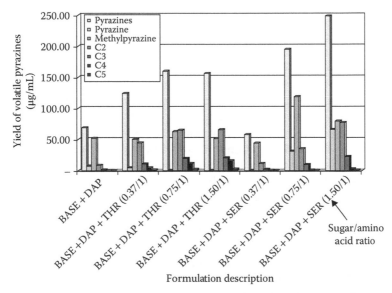

Figure 7.2 Yield of volatile pyrazines in selected microwave heat-treated formulations.

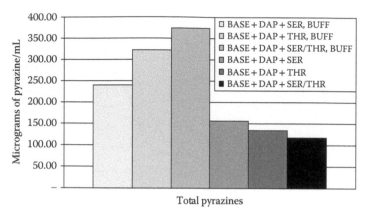

Figure 7.3 Total pyrazines in buffered (BUFF) and unbuffered heat-treated formulations.

2-Methylthiazole

Structure 7.2 General structure for thiazoles.

cases containing THR, the yield of total pyrazines was increased by a factor of two.

Thiazoles, containing a five-member ring with S and N atoms, as a chemical class have been reported to present unique aroma and flavor characteristics. Sulfur-containing amino acids such as cystiene (CYS) in combination with sugars such as glucose have been shown to yield multiple thiazoles (Vernin 1982; Structure 7.2). A brief series of microwave heat treatments on the formulations was made wherein the formulations were fortified with CYS and methionine (MET). Comparison of the headspace volatiles of CYS- and MET-containing formulations revealed the presence of multiple substituted thiazoles in the CYS formulation but not in the MET-containing formulation. Thus, the nature of S bonds in a potential precursor significantly influences the nature of the S-containing compounds produced. The thiazoles found in the CYS-containing formulation (Table 7.4) were consistent with those documented in the literature. These results with CYS confirmed that additional flexibility was available with which to produce more diverse aqueous formulations with unique sensory attributes.

Even though pyrazines, aldehydes, and sugar thermal degradation compounds have dominated the headspace and AUTOSPME profiles of heat-treated formulations to this point, several other compounds with powerful sensory attributes have been found to supplement the profiles. The auxiliary compounds identified in these formulations include substituted pyridines and oxazolines (Structure 7.3). These types of compounds were found as products in sugar/amine reactions (Vernin 1982).

For example, in the heat-treated formulations containing LEU, 2-isobutyl-4, 5-dimethyl-3-oxazoline and 2-(2-methylpropyl)-3,

Table 7.4 Thiazoles Detected in a Base + DAP + CYS
Heat-Treated Formulation

2-Methylthiazole
4-Methylthiazole
3-MethylISOthiazole
5-Methylthiazole
2,4-Dimethylthiazole
2-Ethylthiazole
4,5-Dimethylthiazole
3,4-Dimethylisothiazole
2,4,5-Trimethylthiazole
5-Ethyl-2-methylthiazole
5-Ethyl-4-methylthiazole
4-Ethyl-2,5-dimethylthiazole
2-Acetylthiazole
5-Ethyl-2,4-dimethylthiazole

5-di(1-methylethyl)pyridine were identified. In the VAL and isoleucine (ILE) formulations, isopropyl-substituted oxazolines were detected as well. In a number of other cases, less complex substituted pyridines, such as pyridine and methylpyridine, were detected. As mentioned in the "Introduction," Strecker aldehydes were also present in the headspace

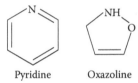

Pyridine   Oxazoline

Structure 7.3 Structures for pyridine and oxazoline.

above these formulations. The Strecker aldehydes 2-methylpropanal, 3-methylbutanal, 2-methylbutanal, benzaldehyde, and phenylacetaldehyde were particularly prevalent in the headspace profiles of formulations when their precursor amino acids, VAL, LEU, ILE, phenylglycine (PHG), and phenylalanine (PHE), respectively, were employed. Thus, the profile of the volatile components in these heat-treated formulations contained not only pyrazines and sugar thermal degradation products (sugar thermals, such as furfural, 5-methylfurfural, and 5-hydroxymethylfurfural) but also relatively small amounts of other volatile components having their own sensory attributes.

The type of HFCS employed in heat-treated formulations was found to have a significant impact on the yield and distribution of selected volatile pyrazines and furfurals. When using HFCS2690, which contains approximately 70% fructose by weight, significantly more pyrazines were generated than when using HFCS55, which contains approximately 42% fructose by weight. By adjusting the mole ratio of sugar to ammonium hydroxide to 1:1, the yield of volatile pyrazines was found to be approximately 4500 µg/mL (Figure 7.4). Yields of sugar thermals including

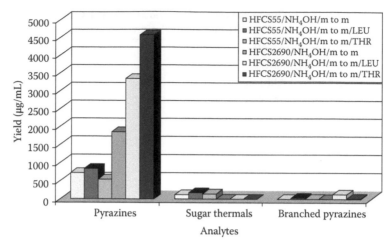

Figure 7.4 Yield of pyrazines and sugar thermals for selected heat-treated formulations.

furfural were found to be higher in cases where HFCS55 was employed (Figure 7.4). With HFCS2690, a pronounced amino acid impact with LEU and THR resulting in notable increases in pyrazine yields was observed.

In addition, through the incorporation of the amino acid LEU the yield of much less volatile branched chain pyrazines was observed. For example, the yield of methyl-3-methylbutylpyrazine was found to be in excess of the yield of sugar thermal degradation products (Figure 7.4) represented by furfural and 5-methylfurfural. Self-aldol condensation products of 3-methylbutanal and 5-methyl-2-isopropyl-2-hexenal, both geometrical *cis* and *trans* isomers, were also detected in these formulations.

Further research was done to more clearly understand the parameters associated with the synthesis of self-aldol condensation products in heat-treated formulations. Formulations rich in branched alkyl chain pyrazines and aldol condensation products could be prepared from formulations containing HFCS, asparagine (ASN), isovaleraldehyde (ISOVAL), and additional nitrogen compounds such as ammonium hydroxide and DAP. The preparation of this particular formulation was performed in a 50-gal scale in a sealed reactor. The relatively low temperature of 220°F of this reaction coupled with short reaction times, less than 5 min, precluded the formation of sugar thermal degradation products such as furfural and 5-methylfurfural. Aldol condensation products resulting from the self-aldol condensation of ISOVAL to yield two hexenal isomers were confirmed. The results obtained from the 50-gal reaction were very similar to those obtained on the laboratory scale in the microwave oven, ~25 mL. In addition to the self-aldol condensation products, a new volatile compound 1-butamine-3-methyl-*N*-(3-methylbutylidene) was detected in this particular formulation

(Structure 7.4). The source of this compound can possibly be attributed to the Schiff base condensation reaction between (3-methylbutyl) amine and 3-methylbutanal.

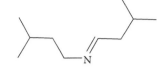

1-Butamine-3-methyl-*N*-(3-methylbutylidene)

Structure 7.4 A new volatile compound.

To discover the optimum level of amino acid to be used in the preparation of pyrazine-rich aqueous formulations, a study was conducted in which different amounts of a selected amino acid were used. The experimental design shown in Table 7.5 was followed, and pyrazines containing branched alkyl substituents, listed in Table 7.6, were identified in the formulations. A typical total ion chromatogram of the headspace volatiles from this set of formulations appears in Figure 7.5 with only the major components labeled.

Stepwise addition of small amounts of amino acid did reveal a point at which maximum yield of pyrazines was obtained, as shown in Figure 7.6. The data plotted in Figure 7.7 is an expansion of the data appearing in Figure 7.6, but in Figure 7.7 the percentage distribution of branched alkyl pyrazines can be clearly seen as a function of the added amino acid level. The percentage of pyrazines attributable to the branched alkyl-containing pyrazines increased dramatically from a low of ~7% to a plateau of ~30% when 1.5 g of LEU was added.

At this point in the understanding of reactions governing yield in aqueous formulations, a body of data was compiled for reactions yielding critical information and findings on parameters critical to the consistent production of formulations rich in pyrazines. These parameters are as follows: (1) reaction temperature, (2) reaction time, (3) reagent concentration, (4) reagent structure, (5) reagent ratios, and (6) pH. Careful control and selection of these parameters would yield formulations that are consistent in content. In addition, careful selection of reagents and reagent ratios afford the capability to produce formulations having the desired complement of pyrazines.

A critical element in the successful understanding of the applicability of laboratory-scale formulations is the ability to transition the laboratory-scale findings to a larger production scale. A number of trials with various formulations and reaction parameters were performed on a 2-gal scale and the characteristics of the headspace volatiles above these formulations were determined. The findings confirmed that formulations rich in branched alkyl chain pyrazines can be prepared from aqueous solutions containing HFCS, selected amino acids, and additional nitrogen compounds such as ammonium hydroxide and DAP. The relatively low temperature of 250°F coupled with long reaction times and the relatively slow ammonium hydroxide addition time of 60 min ensure formulations with rich aromas and relatively high concentrations

Table 7.5 An Exemplary Experimental Design for a Selected Series of Microwave-Prepared Formulations with Selected Amino Acids

Run	HFCS55 (g)	30% NH_4OH (g)	H_2O (g)	LEU (g)	ILE (g)	VAL (g)	Time (min)	Temp (°C)	NH_4OH (moles)	Fructose (moles)	Ratio NH_4OH/FRU
1	15	8.1	6.9	0.5	0	0	30	140	0.07	0.03	2.08
2	15	8.1	6.9	1.0	0	0	30	140	0.07	0.03	2.08
3	15	8.1	6.9	1.5	0	0	30	140	0.07	0.03	2.08
4	15	8.1	6.9	2	0	0	30	140	0.07	0.03	2.08
5	15	8.1	6.9	0	0.5	0	30	140	0.07	0.03	2.08
6	15	8.1	6.9	0	1.0	0	30	140	0.07	0.03	2.08
7	15	8.1	6.9	0	1.5	0	30	140	0.07	0.03	2.08
8	15	8.1	6.9	0	2	0	30	140	0.07	0.03	2.08
9	15	8.1	6.9	0	0	0.5	30	140	0.07	0.03	2.08
10	15	8.1	6.9	0	0	1.0	30	140	0.07	0.03	2.08
11	15	8.1	6.9	0	0	1.5	30	140	0.07	0.03	2.08
12	15	8.1	6.9	0	0	2	30	140	0.07	0.03	2.08
13	15	8.1	6.9	0.5	0.5	0.5	30	140	0.07	0.03	2.08

Table 7.6 Selected Branched-Chain Pyrazines
Detected in Heated Formulations

Pyrazine structure
2-Methyl-5-propylpyrazine
5-Methyl-6,7-dihydrocyclopentapyrazine, isomer
2-Methyl-6-*trans*-propenylpyrazine
5-Methyl-2-isopropyl-2-hexanal, isomer
5-Methyl-2-isopropyl-2-hexanal, isomer
2-Methyl-3-(2-methylpropyl)pyrazine
5-Methyl-3-(2-methylpropyl)pyrazine (tentative)
5-Methyl-6,7-dihydrocyclopentpyrazine, isomer
3-Methylbutylpyrazine
2-Methylpropyl-dimethyl-pyrazine isomer
2-Methylpropyl-dimethyl-pyrazine isomer
2-Methylbutyl-methyl-pyrazine isomer
2-Methylbutyl-6-methylpyrazine
2-Methylbutyl-5-methylpyrazine (tentative)
2,5-Dimethyl-3-methylbutylpyrazine
Dimethyl-3-methylbutylpyrazine isomer
2,6-Dimethyl-3-methylbutylpyrazine (tentative)
Dimethyl-3-methylbutylpyrazine isomer
Dimethyl-3-methylbutylpyrazine isomer
Dimethyl-3-methylbutylpyrazine isomer
Trimethyl-3-methylbutylpyrazine isomer
Trimethyl-3-methylbutylpyrazine isomer
Trimethyl-3-methylbutylpyrazine isomer
Vinyl-(3-methylbutyl) pyrazine (tentative)
2,5-Dimethyl-3,6-(3-methylbutyl) pyrazine (tentative)

of alkyl pyrazines having straight and branched chains. The amount and type of branched-chain pyrazines obtained was a function of the amount and type of the amino acid employed. Precursors associated with pyrazine formation were identified in the formulations. No measurable changes occurred in the headspace of the samples when stored at room temperature for 7 days. The results obtained in these 2-gal reactions were very similar qualitatively and semiquantitatively to those obtained from similar formulations prepared in a microwave reactor on a much smaller scale.

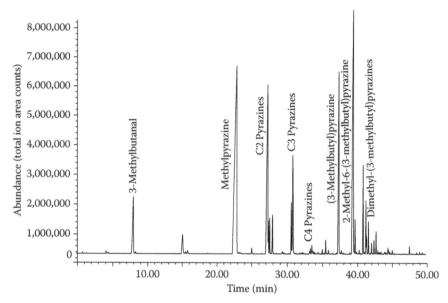

Figure 7.5 Total ion chromatogram of headspace above selected heat-treated formulations. (From Coleman III, W. M., and T. J. Steichen: Sugar and selected amino acid influences on the structure of pyrazines in microwave heat-treated formulations. *J Sci Food and Agric.* 2006. 86:380–391. Copyright Wiley-VCH Verlag GmbH & Co. KGaA. Reproduced with permission.)

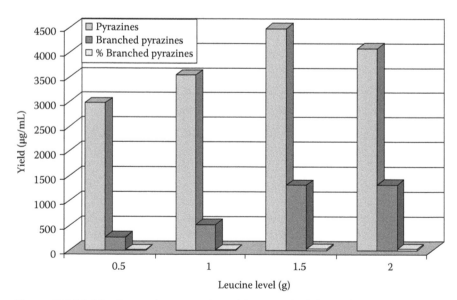

Figure 7.6 Yield of pyrazines in selected heat-treated formulations containing leucine.

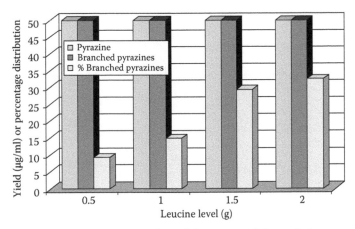

Figure 7.7 Yield of pyrazines in selected heat-treated formulations containing leucine.

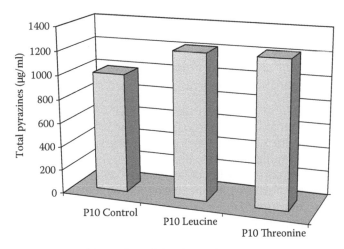

Figure 7.8 Pyrazine yield in selected 2-gal reactions.

Examples of the yields of pyrazines obtained from selected for-mulations prepared in the 2-gal reactor (identified as P10 reactions) are illustrated in Figure 7.8. An additional concern in the preparation of production-scale formulations was the shelf life of material. Thus, head-space volatiles of a selected formulation were measured days after initial preparation (Figure 7.9). The data clearly reveal these formulations were stable for at least a week at room temperature. Such stability is a very posi-tive attribute from a manufacturing/large-scale perspective.

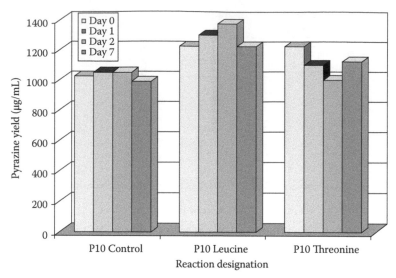

Figure 7.9 Headspace pyrazines in selected 2-gal reactions aged at room temperature.

Initial quantitative analyses (pyrazine standard) and relative quantitative analyses (cyclohexanone internal standard)

Results obtained from the headspace/GC/MSD/FID and the SPME/GC/MSD approaches provide excellent qualitative and relative quantitative information on the nature of the volatile and semivolatile components in several heated aqueous formulations. Conversely, it is reasonably expected that the headspace approach is somewhat deficient in its ability to provide complete qualitative and quantitative information on all the volatile and semivolatile components in such formulations. That is, the headspace approach will most likely not qualitatively or quantitatively detect the presence of relatively semivolatile (less volatile) components in formulations. In addition, components that possess excellent water solubility may not be susceptible to being stripped from the water matrix by the sweep gas or be effectively sorbed onto the SPME fiber. Thus, to gain additional and possibly more-complete information on the nature of compounds not susceptible to analysis by these approaches, 1-µL direct injection of the filtered samples (diluted slightly with methanol) into an HP 5890GC equipped with an HP 5970 MSD were performed. The sample preparation for the GC/MSD analysis was straightforward. Just 1 mL of a selected formulation was diluted with 3 mL of methanol containing the internal standard ethylpyrazine at a known concentration. The highly colored methanol–water solution was placed in a GC vial and injections via an autosampler were made. The software was then applied to the responses obtained to compute quantitative values for the pyrazines.

As predicted, *vide infra*, the level of pyrazines detected by the quantitative GC/MSD approach, ~4500 µg/mL, was much higher than the amount detected by the headspace approach, ~1200 µg/mL, *vide supra*. Furthermore, the percentage contribution of methyl-2-pyrazinylmethanol, undetected by the headspace approach, to the total pyrazines as measured by the GC/MSD approach averaged 42%. The data could also indicate that the presence of amino acids produced formulations having similar total pyrazine levels (Figure 7.10). However, *vide supra*, the amino acids do perform an important role in the production of branched-chain pyrazines that possess very powerful sensory attributes at relatively low concentrations.

In this approach, the entire contents of the formulation are injected into the instrument. Therefore, this technique can be seen to provide for a more complete qualitative and quantitative perspective of volatile and semivolatile compounds in the formulations. A plot of the total ion chromatogram from the direct injection of a selected formulation can be found in Figure 7.11. The results from the GC/MSD approach, as revealed in the distribution of volatile and semivolatile compounds, were substantially different from those found from the headspace approach. Differences between the two results were both qualitative and quantitative in nature. For example, certain compounds identified in Figure 7.11 such as acetic acid and the pyrazinylmethanol derivative were not present in the headspace profiles. Branched-chain pyrazines, which were present

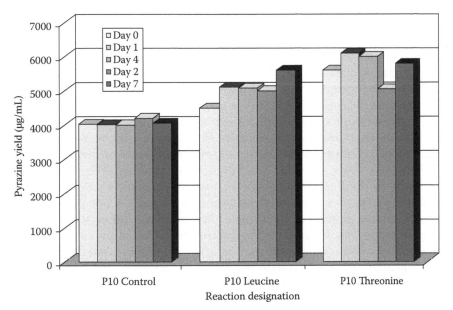

Figure 7.10 Yield of pyrazines from GC/MSD analyses of selected 2-gal reactions aged at room temperature.

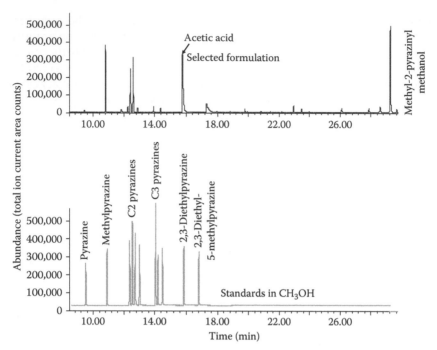

Figure 7.11 Total ion chromatogram from GC–MS analyses of selected samples.

in significant amounts in the headspace approach, were below detection limits in the GC/MSD approach. However, some qualitative similarities can be noted between the two approaches: specifically linear alkyl chain pyrazines such as methylpyrazine were found with both approaches. From a relative quantitative perspective, the dominant pyrazines were the low-molecular-weight linear pyrazines and the compound methyl-2-pyrazinylmethanol (Figure 7.11). This is to be contrasted with the failure of the headspace approach to divulge the presence of the pyrazinylmethanol derivatives. This trend was also evident in studies of microwave-prepared formulations. The proposed relative excellent solubility of the pyrazinylmethanol derivatives in water could explain the inability of the headspace approach to divulge the presence of these pyrazines. From a sensory attribute perspective, pyrazinylmethanol pyrazines do not possess the sensory/aroma impact of pyrazines with branched alkyl substituents. Thus, attempts that focused on controlling the yields of pyrazinylmethanol pyrazines were not as important as attempts that focused on tailoring the yield of branched alkyl chain pyrazines. Nonetheless, these pyrazinylmethanol derivatives could serve as marker compounds for assessing the consistency of batch preparations of these formulations.

Calibration curves were constructed from MSD total ion chromatogram area counts, and quantitative analyses were then performed on these formulations. For the alkylpyrazines, authentic standards were employed to obtain the appropriate response factors. For the pyrazinylmethanol derivative, the response factor attributable to the internal standard, cyclohexanone, was employed to calculate concentration.

To test the ability to produce relatively large-scale quantities of pyrazine-rich formulations, reactions in a 50-gal vessel were attempted. In general, the yield of pyrazines as measured by the headspace technique was found to be in the range of ~450 μg/mL. Numerous additional experiments, however, indicated that the 50-gal preparation would produce formulations with slightly lower pyrazine yields, ~400 μg/mL, when compared with the microwave reactor results, 450 μg/mL. However, the scale-up procedures produced formulations that were very similar in qualitative composition to formulations from laboratory experiments. A more advanced quantitative analysis of pyrazine formulations involving SIM MSD appears in section titled *Quantitative gas chromatography/selected ion monitoring–mass selective detection analyses employing multiple authentic pyrazine standards.*

Model reactions

During the course of investigations into pyrazine yield optimization based in part on formulations containing HFCS, numerous examples of compounds attributable to sugar thermal degradation processes were observed. Specifically, compounds such as 3-hydroxy-2-propanone and 3-hydroxy-2-butanone appear in the headspace above the formulations. Literature references postulate the presence of such molecules in mechanistic arguments for the production of pyrazines in model systems (Shibamoto and Bernhard 1977; Yaylyan and Keyhani 1999; Chen and Ho 1998). In addition, Strecker aldehydes have been postulated as key intermediates in the production of pyrazines (Chen and Ho 1998). Thus, a series of experiments was conducted to examine the possibility and viability of producing formulations rich in pyrazines from alternative carbon sources in combination with Strecker aldehydes.

Commercially available HFCS is a mixture of water, glucose, and fructose having a well-defined composition. In experiments with alternative carbon sources, aqueous solutions were prepared having concentrations of 3-hydroxy-2-propanone and dihydroxyacetone that were equivalent, on a mole-to-mole basis, to the concentration of fructose in HFCS-containing aqueous formulations. For example, a series of formulations containing a set amount of 3-hydroxy-2-propanone (0.028 moles), a set amount of NH_4OH (0.028 moles), a set amount of water, and varying amounts of a selected Strecker aldehyde was heat treated in a microwave oven at 105°C for 60 min. The headspace above these heated formulations was measured and the results are presented in Figure 7.12. It is important to note that headspace volatile yields were calculated based on flame ionization

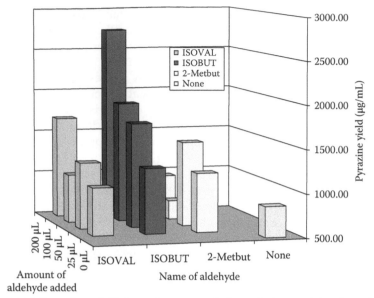

Figure 7.12 Pyrazine yield from heat-treated formulations with selected aldehydes. ISOVAL = isovaleraldehyde, ISOBUT = isobutyraldehyde, and 2-Metbut =
2-methylbutyraldehyde.

detector response to a known amount of the internal standard, that is,
cyclohexanone. Thus, these values were described as being relative quantitative in nature. However, these data still allowed direct comparisons
of pyrazine yields as a function of defined variables. The mass selective
detector was used to confirm the structure of the volatile components.
Several important points were apparent from the data: The yield of pyrazines was greater than or equal to 1000 μg/mL, the yield of pyrazines
was a function of the amount of aldehyde present, the aldehyde structure had an impact on the yield of pyrazines, and the amount of aldehyde
influenced the total yield of pyrazines.

One of the primary motivations for including aldehydes in the formulation was to study the potential for producing heat-treated formulations
with pyrazines having branched alkyl substituents attached to them.
In these experiments, the percentage yield of pyrazines with branched
alkyl substituents was calculated by dividing the yield of these pyrazines
by the total yield of pyrazines. The data plotted in Figure 7.13 convincingly demonstrate that inclusion of aldehydes altered the percentage of
pyrazines with branched alkyl substituents (branched pyrazines) in the
formulations. For example, with no aldehyde present the percentage of
branched pyrazines was approximately 8%. Conversely, with as little as
25 μL (~0.0002 moles) of added aldehyde, the percentage of branched pyrazines increased from ~25%. It was noted that stepwise increases in the

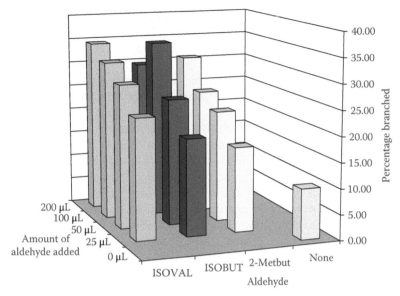

Figure 7.13 Percentage of branched alkyl pyrazine yield from heat-treated formulations with selected aldehydes.

(3-Methylbutyl)pyrazine   (*cis*-2-Propenyl)pyrazine

Structure 7.5   **Structure 7.6**

amounts of all the aldehydes produced similar stepwise increases in the percentage of pyrazines containing branched alkyl substituents.

From a qualitative perspective, the branched pyrazines in these formulations were, as expected, a function of the structure of the aldehyde used. For example, among the dominant branched pyrazines for formulations containing ISOVAL, isobutyraldehyde (ISOBUT), and 2-methylbutanal, there were (3-methylbutyl)pyrazines, (2-methylbutyl) pyrazines, and (2-methylpropyl)pyrazines, respectively (Structure 7.5). Additionally, other types of branched pyrazines were noted in these particular compositions (Structure 7.6). Specifically, *cis* and *trans* propenyl pyrazines were identified and noted in formulations containing 3-hydroxy-2-propanone (3-OH-2-propanone).

By way of comparison, a series of microwave reactions was performed wherein the sole change was substitution of HFCS for 3-OH-2-propanone on a mole-to-mole basis. This mole-to-mole substitution was based on the amount of fructose present in the HFCS. Results indicated the yield from HFCS to be approximately 1300 µg/mL compared with approximately 3000 µg/mL from the combination of aldehydes with 3-OH-2-propanone.

These trends seem to be consistent in light of the mechanisms postulated about the role that fructose plays in the formation of pyrazines, that is, the production of 3-OH-2-propanone. Use of 3-OH-2-propanone instead of fructose from HFCS demonstrated at least two trends: (1) elimination of the fructose "decomposition" step and (2) enhancement of the concentration of a proposed key reaction intermediate. Thus, selected water-soluble organic compounds, postulated intermediates from fructose decomposition, could more than compensate for the use of HFCS as a substrate for the production of cooked formulations. Furthermore, these molecules yielded compositions having superior levels of pyrazines and branched pyrazines.

Selected amino acids have been employed very effectively in heat-treated aqueous formulations for their capability to yield enhanced levels of branched pyrazines, *vide supra* (Hwang, Hartman, and Ho 1995; Coleman 1999a,b). Thus, the potential for the use of amino acids in concert with alternative carbon sources seem worthy of investigation. To gain some insight into this proposal, a series of formulations was prepared employing selected amino acids at a constant concentration, with both HFCS and 3-OH-2-propanone as carbon sources. Numerous significant observations could be made regarding the performance of amino acids in both HFCS and 3-OH-2-propanone formulations:

- Regardless of amino acid employed, the yield of pyrazines was highest when 3-OH-2-propanone was used.
- The amount of branched pyrazines was always highest in compositions containing 3-OH-2-propanone.
- Strecker aldehyde-forming amino acids yielded the most pyrazines.
- In general, the yield of pyrazines and branched pyrazines increased when amino acids were employed.
- Amino acid-containing formulations were much less effective (on a mole-to-mole basis) in producing branched pyrazines than comparable Strecker aldehydes.

In the amino acid cases, an intermediate reaction, known as the amino acid Strecker degradation, must occur prior to making a Strecker aldehyde available for further reaction in the formation of branched pyrazines. The use of the pure Strecker aldehyde eliminated the need for an intermediate Strecker degradation reaction. In the same manner, the use of 3-OH-2-propanone or dihydroxyacetone instead of HFCS (fructose) eliminated the need for an intermediate sugar degradation reaction.

During the course of investigations directed at the optimization of pyrazine-rich formulations, addition of DAP was observed to decrease the overall yield of pyrazines, although DAP has ammonium ions as part of its structure. Several possibilities for the impact of DAP were put forward. But it was not until a series of experiments was conducted using sugar phosphate esters that a very plausible mechanistic explanation was arrived at. Such an explanation could include a new proposal versus the one based on pure pH influence, *vide supra*. Aqueous solutions containing equimolar amounts of selected sugar phosphate esters along with fructose and glucose were heat treated in a microwave oven at 120°C for 30 min. The amounts of pyrazines were measured in the headspace of each reaction formulation and the results are shown in Figure 7.14. The results of these experiments with sugar phosphates and with the addition of DAP to aqueous solutions of glucose and fructose strongly indicated the formation of sugar esters in the presence of DAP. Once the esters were formed, the yield of pyrazines from such sugar esters was substantially reduced.

The distribution of pyrazines resulting from the preparation of these types of formulations has long been recognized as an important link to the sensory attributes of such formulations. This is particularly evident from pyrazine structure–activity relationships odor threshold data available in the open literature (Buchbauer et al. 2000; Wagner et al. 1999; Fors 1983). In particular, findings in the literature are consistent in that pyrazines with four carbons attached, for example, ethyldimethylpyrazines,

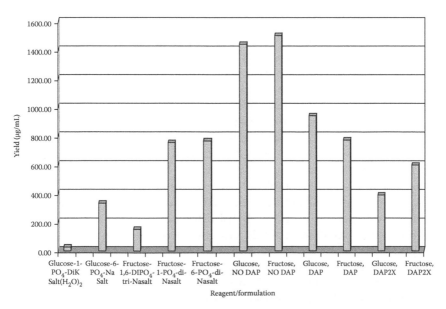

Figure 7.14 Yield of headspace pyrazines in selected heat-treated formulations.

have the lowest odor thresholds of any of the low-molecular-weight pyr-azines. The relative differences in odor thresholds could be illustrated by assigning methylpyrazine an impact factor of 1, and the resulting impact factor for ethyldimethylpyrazines would be approximately 125,000. Thus, further optimization of these formulations focus in part on the prepara-tion of formulations rich in pyrazines with four carbons attached, or C4 pyrazines. A detailed close review of the literature reveals that rhamnose, used as a sugar source in model studies with several sugars, was the only sugar that produced measurable quantities of C4 pyrazines (Shibamoto and Bernhard 1977). More specifically, when sugars and ammonium hydroxide were reacted at 100°C for 2 h in an aqueous medium, measur-able amounts of pyrazines were produced. These reaction conditions were immediately recognized as being well within the range of parameters used in the preparation of these formulations. Thus, investigations into the characteristics of formulations prepared with rhamnose as the sugar source were initiated.

Selection of sugar and amino acid was again shown to dictate the structure, yield, and distribution of headspace-volatile pyrazines. Under the same reaction conditions and sugar-to-ammonium hydroxide ratios (Shibamoto and Bernhard 1977), rhamnose was found to produce more pyrazines than HFCS. With rhamnose, the yield of pyrazines approached 4500 μg/mL. Formulations selectively enriched in ethyldimethylpyr-azines can be prepared by stepwise substitution of HFCS with rhamnose. Formulations preferably enriched in methyl- and dimethyl-(3-methylbutyl) pyrazines as well as methyl- and dimethyl-(2-methylpropyl)pyrazines were prepared through selective combinations of rhamnose/HFCS and amino acids. Stepwise substitution of HFCS with rhamnose coupled with the use of selected amino acids allowed for the control of yield, structure, and distribution of volatile pyrazines in microwave heat-treated aqueous formulations. For example, a comparison of headspace pyrazine yields in microwave heat-treated formulations with rhamnose and HFCS indicated a temperature and sugar-to-nitrogen ratio (S/N) influence on the yield of pyrazines (Figure 7.15). Accompanying the influence of S/N and tem-perature there was a shift in the distribution of pyrazines (Figure 7.16). The impact of rhamnose incorporation into the formulations in terms of the distribution of low-molecular-weight pyrazines was immediately evident. Systematic substitution of rhamnose for HFCS precipitated a dramatic systematic shift in the distribution of the pyrazines. With 100% rhamnose, the contribution of C4 pyrazines was almost as great as that of C1 pyrazines.

Having shown earlier that rhamnose was capable of altering the distribution of methyl, C2, C3, and C4 pyrazines, would rhamnose-containing formulations fortified with amino acids produce further altered distributions of pyrazines with branched alkyl chains?

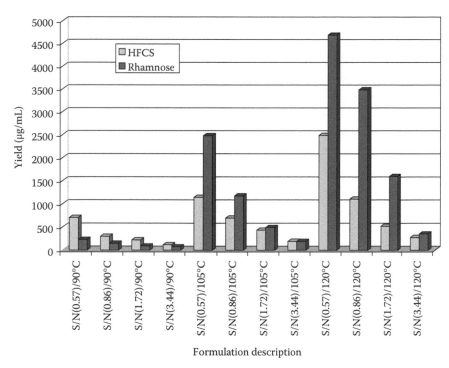

Formulation description

Figure 7.15 Comparison of headspace pyrazine yields in selected heat-treated formulations with rhamnose and high-fructose corn syrup. (From Coleman III, W. M., and T. J. Steichen: Sugar and selected amino acid influences on the structure of pyrazines in microwave heat-treated formulations. *J Sci Food Agric.* 2006. 86:380–391. Copyright Wiley-VCH Verlag GmbH & Co. KGaA. Reproduced with permission.)

Under all the microwave reaction conditions and reagent ratios examined in the aforementioned studies, formulations containing only rhamnose produced higher percentages of branched alkyl chain pyrazines than did formulations containing only HFCS. In some cases, the branched alkyl chain pyrazine values for formulations containing rhamnose were approximately twice the values for formulations containing HFCS. Additional information on the role of rhamnose in producing changes in branched alkyl chain pyrazines can be found in the yields of specific pyrazines within the general category of branched alkyl chain pyrazines. For example, based on the established role of rhamnose, increases in the yields of branched alkyl chain pyrazines with accompanying methyl and dimethyl substituents might be expected when rhamnose is employed. Using 2,5-dimethyl-(3-methylbutyl)pyrazine as a representative of pyrazines containing both branched and linear alkyl chains, the influence of rhamnose on the yield of these molecules versus the influence of HFCS was scrutinized. At 105°C and 120°C, the yield of 2,5-dimethyl-(3-methylbutyl)

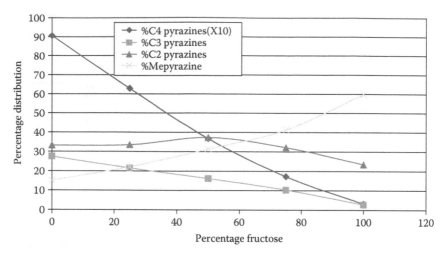

Figure 7.16 Percentage distribution of headspace pyrazines in selected heat-treated formulations (120°C/60 min) with a sugar-to-nitrogen ratio of 0.57:1.00. (From Coleman III, W. M., and T. J. Steichen: Sugar and selected amino acid influences on the structure of pyrazines in microwave heat-treated formulations. *J Sci Food Agric.* 2006. 86:380–391. Copyright Wiley-VCH Verlag GmbH & Co. KGaA. Reproduced with permission.)

pyrazine was much higher in rhamnose-containing formulations than in the comparable HFCS cases. In some cases, the yield was four times that obtained with HFCS. Thus, stepwise substitution of HFCS with rhamnose coupled with the use of selected amino acids, LEU and VAL, allowed for the control of yield, structure, and distribution of volatile low-molecular-weight pyrazines in microwave heat-treated formulations.

To this point in the development of these formulations, the dominant form of reagent preparation involved the addition of aqueous ammonium hydroxide to an aqueous solution of a selected sugar and/or amino acid. The outcome of adding solid sugar to a concentrated ammonium hydroxide/amino acid solution followed by microwave heat treatment had yet to be examined. Thus, a sequence of experiments was designed to discover the effects of this alternative addition procedure. Microwave heat treatment of solutions at 110°C for 1 h produced formulations rich in pyrazines with pyrazine levels greater than 4000 μg/mL. These yields were significantly higher than those currently obtained employing HFCS with dropwise addition of concentrated ammonium hydroxide under the same process conditions (Figure 7.17). Stepwise decrease in fructose in the fructose–rhamnose sugar mixture resulted in increased pyrazine yields (Figure 7.17). However, when the sugar/ammonium hydroxide solutions were aged overnight, followed by microwave heat treatment, pyrazine concentrations were found to be lower than those found in nonaged

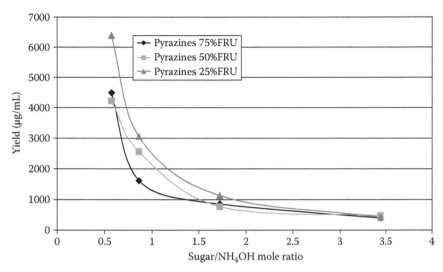

Figure 7.17 Yield of headspace pyrazines as a function of sugar structure and ammonium hydroxide (NH₄OH) content.

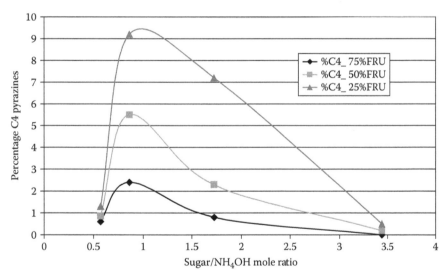

Figure 7.18 Percentage of C4 pyrazine as a function of sugar source and ammonium hydroxide (NH₄OH) concentration.

formulations. Formulations containing only rhamnose produced approximately 2× the amount of pyrazines, ~10,000 µg/mL, produced by fructose. Rhamnose-containing formulations consistently produced formulations having greater than 10% C4 pyrazines. Substitution of rhamnose with fructose resulted in a decrease in C4 pyrazine contribution (Figure 7.18).

A strong influence of sugar:NH$_4$OH ratio was evident in the C4 production rate (Figure 7.18), indicating a kinetic influence. Addition of amino acids, LEU and VAL, to the formulations resulted in materials consistently having greater than 15% branched alkyl chain pyrazines.

The literature also indicates that addition of the amino acid alanine (ALA) to a formulation results in increases in the percentage of C4 pyrazines (Hemaini, Cerny, and Fay 1995). Addition of ALA to formulations containing rhamnose as well as formulations containing HFCS boosted the percentage of headspace C4 pyrazines. In the rhamnose cases, the increase was from ~8% to ~12%, whereas for HFCS-containing formulations the increase was from ~0.4% to 2.0%. Formulations containing LEU, VAL, and ALA produced significantly more headspace-volatile branched alkyl chain pyrazines than comparable formulations containing no LEU and VAL. Increasing levels of ammonium hydroxide were found to produce increasing yields of headspace-volatile pyrazines.

Quantitative gas chromatography/selected ion monitoring–mass selective detection analyses employing multiple authentic pyrazine standards

During the course of investigations, a purposeful, systematic move toward more quantitative assessments of aqueous formulations was made. More specifically, a method based on GC/SIM-MSD was developed. Use of authentic standards for pyrazines allowed quantitative analysis of formulations. Use of the SIM mode assisted in quantifying the pyrazines at lower detection limits. Quantitative analysis using GC/SIM-MSD confirmed the relative quantitative trends observed with headspace experiments. For example, the addition of ALA to a formulation did result in a measurable increase in the percentage of C4 pyrazines obtained. In addition, the SIM-based quantitative approach revealed approximately two-to-three times the concentration of pyrazines that were previously revealed by the relative quantitative headspace approach. This was not surprising as the internal standard, cyclohexanone, employed for headspace analyses does not resemble pyrazines in structure. However, the distinct advantage of cyclohexanone is that in none of the experiments was cyclohexanone observed to be produced. Thus, the response from cyclohexanone would remain consistent across many experiments and free from interferences.

Use of the GC/SIM-MSD quantitative approach revealed additional information on the differences obtained through the use of rhamnose (RHAM50, a rhamnose/NH$_4$OH/LEU/VAL formulation) versus HFCS (P13, a HFCS/NH$_4$OH/LEU/VAL formulation; Table 7.7) for reactions performed in a 50-gal reactor. With average percentage RSD values consistently less than 5%, some meaningful information and trends were deduced from the data in Table 7.7:

Table 7.7 A Quantitative Comparison of the Yield and Percentage Distribution of Alkylpyrazines between Two Heat-Treated Formulations

Pyrazine	Formulation (µg/mL)		Formulation (percentage distribution)	
	P13	RHAM50	P13	RHAM50
Methyl	630.37	3537.96	37.08	22.31
2,5-Dimethyl	76.73	870.52	4.51	5.49
2,6-Dimethyl + ethyl	523.71	5868.54	30.81	37.01
2,3-Dimethyl	56.50	505.49	3.32	3.19
2-Ethyl-6-methyl	132.04	2614.34	7.77	16.49
2-Ethyl-5-methyl	ND	381.58	0	2.41
2,3,5-Trimethyl	170.95	1192.48	10.06	7.52
2-Ethyl-3,5-diethyl	3.01	103.55	0.18	0.65
2,3-Diethyl	ND	500.40	0	3.16
2-Ethyl-3,6-dimethyl	3.67	283.12	0.22	1.79
Tetramethyl	2.31	ND	0.14	0
23-Diethyl-5-methyl	ND	ND	0	0

ND = not detected

1. The formulation RHAM50 had a significantly higher concentration of pyrazines than did P13, consistent with the headspace results.
2. The quantitative GC/SIM-MSD approach revealed there to be more actual pyrazines than those found with the relative quantitative headspace approach. For example, the headspace analysis of RHAM50 indicated the pyrazine yield to be approximately 9000 µg/mL.
3. The RHAM50 formulation had a significantly lower concentration of methylpyrazine than did the P13 formulation, consistent with the headspace results.
4. The RHAM50 formulation had a significantly higher concentration of C4 pyrazines, 5.6%, than did the P13 formulation, 0.54%, consistent with the headspace results.

Thus, rhamnose-based heat-treated formulations were shown, on a quantitative basis, to produce more pyrazines and a greater percentage of C4 pyrazines than an HFCS-based formulation.

Open vessel reactions

All the preparations for pyrazine-rich formulations described to this point employed closed pressure vessels to yield formulations that are substantially rich in pyrazines. A preliminary experiment revealed that with rhamnose one can potentially prepare formulations rich in pyrazines without the use

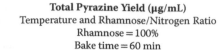

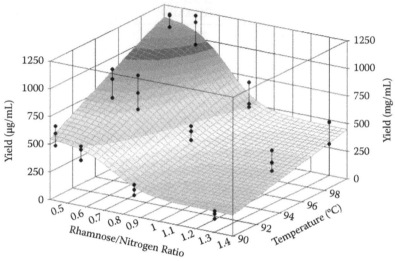

Figure 7.19 Response surface for pyrazine yields versus temperature and rhamnose/nitrogen (ammonium hydroxide [NH₄OH]) ratio. (From Coleman III, W. M., and T. J. Steichen: Sugar and selected amino acid influences on the structure of pyrazines in microwave heat-treated formulations. *J Sci Food Agric.* 2006. 86:380–391. Copyright Wiley-VCH Verlag GmbH & Co. KGaA. Reproduced with permission.)

of a pressure vessel. To further test and confirm this preliminary observation, a sequence of experiments was conducted using a round-bottom flask open to the atmosphere. By using rhamnose coupled with dropwise NH₄OH addition, aqueous formulations rich in not only pyrazines but also caramel flavor compounds could be prepared in an open vessel at approximately 100°C. From reactions performed in a round-bottom flask fitted with a condenser and stirrer, the yields and distributions of pyrazines and caramel compounds were shown to be functions of the reaction time and temperature (Figure 7.19). After approximately 2 h at 100°C, the yield of pyrazines reached a stable plateau at ~900 µg/mL (Figure 7.19). The trends in pyrazine yields were confirmed with both the quantitative GC/SIM-MSD approach and the relative quantitative AUTOSPME/GC/SIM-MSD approach. One of the very positive process advantages demonstrated by the open vessel approach is that the reaction occurred at the boiling point of water, which is relatively easy to control in a process environment.

To more clearly establish the optimum conditions under which to prepare pyrazine-rich formulations in an open vessel, a succession of experiments

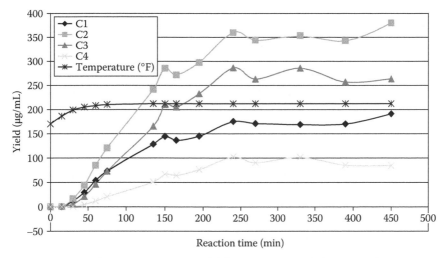

Figure 7.20 Headspace pyrazines yield for a heat-treated rhamnose/ammonium hydroxide (NH_4OH) formulation at 212°F in an open vessel.

was performed wherein temperature and rhamnose-to-nitrogen ratios were varied. Through the use of rhamnose, LEU, VAL, and NH_4OH, aqueous formulations rich in pyrazines could be prepared in an open vessel at approximately 100°C. From reactions in which reagents, time, and temperature were systematically varied, the yields and distributions of pyrazines were shown to be functions of reaction time (20–120 min), temperature (90°C–100°C), and the S/N mole ratio (1.3/1 to 0.12/1). The variations in yield and distribution were found to be nonlinear with respect to S/N with dramatic increases in pyrazine yield being observed for an S/N between 0.86 and 0.43 (Figure 7.19). Moderate systematic increases in yield were observed between 90°C and 100°C with yields being the highest at 100°C. Increases in reaction time produced systematic increases in yield with the maximum occurring at a reaction time of 220 min (Figure 7.20). The data also revealed that C4-pyrazine yields in these formulations followed a similar pattern to that revealed for total pyrazine yield (Figure 7.20). At 88°C, the yield of pyrazines continued to increase even after 7 h. After 7 h at 100°C, the yield of caramel flavor compounds, ~600 µg/mL, also continued to increase. Thus, formulations having ample amounts of two important classes of flavor compounds can be prepared without requiring the use of a pressure vessel.

The concept of open vessel technology for the preparation of pyrazine-rich formulations was extended to include the 2-gal reactor. The 2-gal reactor was fitted with a condenser and the reaction was performed using the reagent ratios and conditions established from smaller-scale laboratory microwave results for maximum pyrazine yield. After completing the addition of ammonium hydroxide, a small amount of condensate was collected from the top of the condenser. This clear, colorless condensate had a very powerful pyrazine aroma. The concentration of pyrazines in this condensate was measured to be greater than 2000 µg/mL. The results obtained from the 2-gal reactions thus reflected those obtained in the laboratory-scale reaction conducted in a 250-mL round-bottom flask. Thus, these laboratory- and pilot-scale experiments reconfirmed that pyrazine-rich aqueous solutions can be effectively distilled from a reaction occurring in an open vessel (Coleman and Gerardi 2010).

There is convincing evidence in the literature that Amadori compounds are involved in the production of volatile compounds when natural products are subjected to heating. Amadori compounds are those species that are formed upon reaction of an amino acid with a sugar. These compounds can be isolated as solid materials, indicating some measurable degree of stability. Thus, it was of interest to examine the types of volatile and semivolatile compounds that would result from the pyrolysis of selected Amadori compounds. From a qualitative perspective, the array of volatiles was shown to be influenced by the structure of the Amadori compound. More specifically, fructose-based Amadori compounds derived from ASN and THR were shown to have a more diverse array of volatile compounds than fructose-based Amadori compounds derived from LEU and VAL. A large part of the additional diversity was attributed to the presence of pyridines and pyrazines. The compounds identified in this study are typical of the types of volatiles found in tobacco and tobacco smoke samples and similar to those found in formulations consisting of HFCS and amino acids. More specifically, those pyrazines previously directly associated with selected amino acids (Structure 7.1) were identified in the pyrolysate of the respective amino acid (Coleman 2002a; Table 7.8). Notice once again how the structure of the detected volatile components is directly linked to the structure of the Amadori compound. Additional work at selected temperatures revealed the presence of intermediates previously only postulated to occur in the reaction between amino acids and sugars and the decomposition of Amadori compounds (Coleman 2002b).

Studies employing aqueous solutions and suspensions containing labeled reagents (sealed vessels)

The volatile and semivolatile compounds detected in heat-treated aqueous formulations using headspace techniques, quantitative target compound analyses, as well as SPME analyses have been shown to be dominated by three general categories of compounds: (1) sugar thermal degradation compounds like furfurals and ketones, (2) S/N reactions producing compounds like pyrazines, and (3) Strecker degradation of Amadori compounds to yield low-molecular-weight aldehydes. These compounds were the dominant types found in the headspace regardless of processing conditions

Table 7.8 Pyrolysis/GC–MS Qualitative Analysis of
Selected Amadori Compounds

	VAL-D-	LEU-D-	ASN-D-	THR-D-
Major compound/compound types	FRU	FRU	FRU	FRU
2-Methylpropanal	x			
3-Methylbutanal		x		
2-Methylpropanoic ACID	x			
3-Methylbutanoic ACID		x		
2-Methyl-3-(2-methylpropyl) pyrazine	x			
6-Methyl-2-(3-methylbutyl)pyrazine		x		
Dimethyl-(2-methylpropyl) pyrazines	x			
Dimethyl-(3-methylbutyl)pyrazines		x		
Trimethyl-(3-methylbutyl)pyrazines		x		

(Structures shown in the ASN-D-/THR-D- columns: Valine — NH_2, OH, O; Leucine — NH_2, OH, O.)

FRU = Fructose

(Coleman 1992; Coleman, White, and Perfetti 1994, 1996; Coleman and Perfetti 1997; Coleman 1997, 1999a; Coleman and Gerardi 2010).

Based in part on the supposition that the free amino acids and ammonium ions present in a natural product, such as cured tobacco, are key reagents in the production of pyrazines, a series of preliminary heat treatment experiments was performed on tobacco suspensions. The results revealed that the role of the amino acids and ammonium ions were clearly defined and expressed in two familiar major pathways: (1) amino acid–Maillard reactions with Strecker degradation and (2) sugar–ammonium ion reactions. The mechanisms of Strecker degradation and pyrazine formation were further elucidated by the selection of a target amino acid with a well-defined volatile aldehyde degradation product such as LEU. For example, the Strecker degradation aldehyde linked with LEU is 3-methylbutanal. Thus, employing ^{13}C LEU and ^{12}C LEU as amino acids in a typical heat-treated formulation should have produced ^{13}C-labeled 3-methylbutanal and ^{12}C-labeled 3-methylbutanal, respectively. The data plotted in Figures 7.21 and 7.22 convincingly confirm these hypotheses. The Strecker aldehyde 3-methylbutanal has been shown to be a dominant volatile material in the heat-treated formulation. The mass spectra (Figure 7.22) confirm the presence of ^{13}C and [12]C 3-methylbutanal molecules.

Strecker aldehydes have also been implicated in playing an essential role in determining the type of alkyl substituents that appear on substituted pyrazines. Thus, the use of ^{13}C and ^{12}C LEU as ingredients could reasonably be predicted to yield ^{13}C and [12]C 3-methylbutylpyrazines. The data plotted in Figures 7.23 and 7.24 confirm this hypothesis. It is obvious that

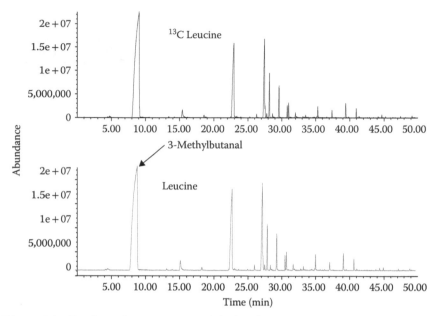

Figure 7.21 Total ion chromatogram of the headspace above a leucine-containing selected heat-treated formulation at 175°C/30 min.

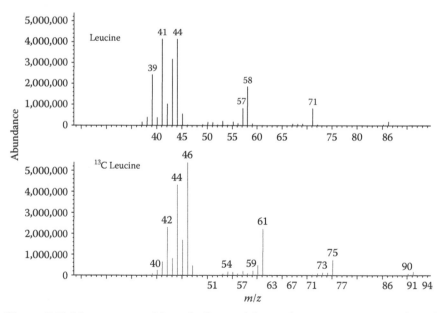

Figure 7.22 Mass spectra of 3-methylbutanal from a leucine-containing selected heat-treated formulation at 175°C/30 min.

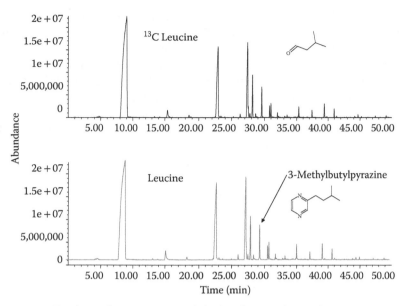

Figure 7.23 Total ion chromatogram of the headspace above a leucine-containing selected heat-treated formulation at 175°C/30 min.

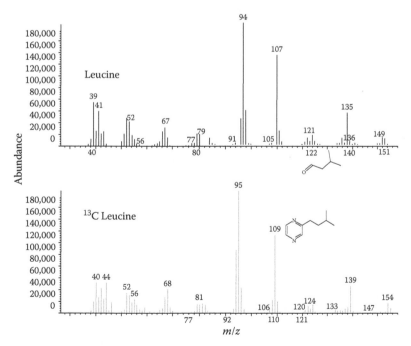

Figure 7.24 Mass spectra of 3-methylbutylpyrazine from a leucine-containing selected heat-treated formulation at 175°C/30 min.

the carbons from the Strecker aldehyde 3-methylbutanal have been incorporated into the pyrazine molecule. Further confirmation can be realized by observing that the m/z ratios vary in concert with an increase of 5 m/z units attributable to the ^{13}C carbons in the labeled Strecker aldehyde.

In a similar fashion as to that employed for Strecker aldehyde participation in pyrazine formation, the mechanisms surrounding amino acid and ammonium ion reactions were substantiated by adding a selected number of uniformly ^{15}N labeled amino acids and ammonium ions. For example, addition of uniformly ^{15}N labeled ammonium acetate ($^{15}NH_4OAc$) to a natural product suspension followed by heat treatment yields methylpyrazine having both its nitrogen atoms labeled as ^{15}N. The data appearing in Figure 7.25 confirm the increase in m/z of 2 units. In a similar set of experiments, the role of amino acids in pyrazine formation was documented in heat-treated licorice formulations (Figure 7.26). By measuring the ratios of m/z abundances, the percentage of ^{15}N incorporation could be determined. The data convincingly showed that urea was very effective as a nitrogen source for pyrazine formation, whereas amino acids and another nitrogen source, ammonium acetate, produced pyrazines having varying levels of ^{15}N incorporation.

At equal molar equivalents, addition of the N^{15}-labeled amino acids and ammonium ions to a tobacco suspension followed by heat treatment precipitated significant increases in the concentration of volatile pyrazines and aldehydes relative to the control tobacco with no added amino acids

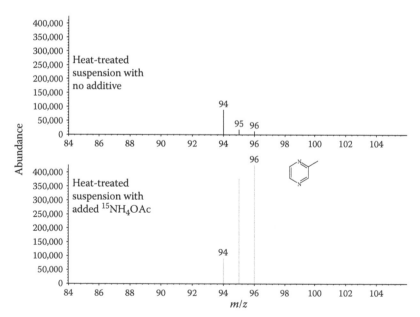

Figure 7.25 Selected ion monitoring mass spectra of methylpyrazine produced in selected heat-treated natural product suspensions.

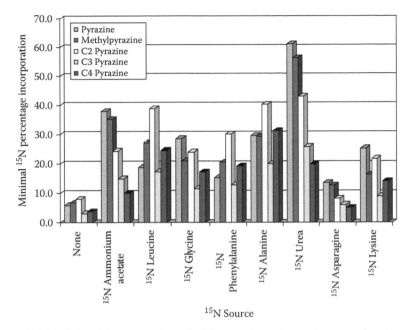

Figure 7.26 Minimal incorporation of ^{15}N into selected pyrazines as a function of added ^{15}N sources in licorice heat-treated formulations.

and ammonium ions (Figure 7.27). When the mass spectra were studied carefully, the volatile pyrazines were found to have ^{15}N atoms distributed throughout the entire pyrazine array. However, on an equal molar basis, the ^{15}N source influenced the type of pyrazines produced. For example, addition of ^{15}N by means of ammonium acetate produced pyrazine with ^{15}N nitrogens, whereas addition of ^{15}N PHE predominately produced ^{15}N-containing pyrazines with four carbons attached (C4 pyrazines, i.e., dimethylethylpyrazines). Similar results were obtained with licorice trials also. Thus, these results provide additional insights into the dominant mechanisms involved in the production of volatile compounds from the heat treatment of a candidate natural product.

Additional investigations were undertaken to document similar reactions in other natural products. These efforts culminated in the discovery that a unique commonality existed between a wide range of natural products in terms of the volatile materials produced upon heat treatment (Coleman 1996). Regardless of the starting natural product, the same qualitative profile of volatile materials was produced upon "pressure cooking"/heat treatment of an aqueous extract of a natural product. The notable differences in aroma detected for the heat-treated natural products were ascribed to significant quantitative and not significant qualitative differences in the extracts. Differences in the free amino acid content of aqueous

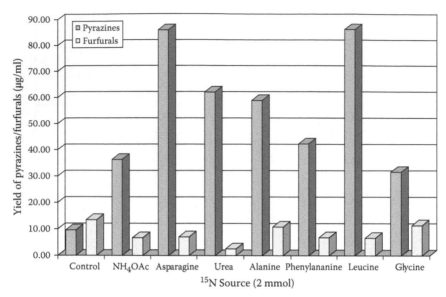

Figure 7.27 Yield of pyrazines and furfurals as a function of added ^{15}N source.

Table 7.9 Typical Formulation Descriptions for
Heat-Treated Tobacco Suspensions

1. Tobacco + no additive
2. Tobacco + 2 mmol ^{15}N LEU + 2 mmol ^{13}C$_6$ sugar
3. Tobacco + 2 mmol ^{15}N glycine + 2 mmol ^{13}C$_6$ sugar
4. Tobacco + 2 mmol ^{15}N PHE + 2 mmol ^{13}C$_6$ sugar
5. Tobacco + 2 mmol ^{15}N ALA + 2 mmol ^{13}C$_6$ sugar
6. Tobacco + 2 mmol ^{15}N ammonium acetate + 2 mmol ^{13}C$_6$ sugar
7. Tobacco + 2 mmol ^{15}N ASN + 2 mmol ^{13}C$_6$ sugar
8. Tobacco + 2 mmol ^{15}N urea + 2 mmol ^{13}C$_6$ sugar

extracts were given as a plausible explanation for the differences found in headspace volatiles. For example, the aqueous extract of alfalfa hay had the highest concentration of ILE, LEU, and VAL of all the products examined, and on heat treatment it yielded the highest levels of Strecker aldehydes, 2-methylpropanal, 3-methylbutanal, and 2-methylbutanal. All the extracts produced volatile reaction products in the three aforementioned categories for tobacco dust.

Based on the results obtained on the natural product array, additional, more detailed studies of the roles of amino acids and sugars in the production of volatile materials in microwave-based heated tobacco suspensions (HTTSs) were undertaken. Typical formulation constituents are listed in Table 7.9. The results of these studies add more clarity to the understanding of roles played by amino acids and sugars in the production of volatiles as a result of heat treatment. The mechanism of Strecker degradation was confirmed by the addition of selected uniformly ^{13}C labeled amino acids to the reaction suspension. The addition of the uniformly ^{13}C labeled amino acids resulted in a dramatic increase in the concentration of specific and appropriate uniformly ^{13}C labeled low-molecular-weight Strecker aldehydes in the headspace above the reaction suspension. The mechanism of the amino acid–sugar Maillard-type reactions was further clarified by adding a wide array of uniformly ^{13}C labeled amino acids to the suspension, employing ^{15}N-labeled amino acids, and employing uniformly ^{13}C labeled sugars. Addition of amino acids to the suspension followed by heat treatment resulted in significant increases (2×) in the concentration of volatile pyrazines relative to the control. Not all amino acids increased the concentration of headspace pyrazines. Of the amino acids added to the suspension, LEU, glycine, PHE, and ALA resulted in pyrazines having as high as 90% of their nitrogens labeled. Reactivity differences as a function of sugars (glucose and fructose) were also demonstrated. For example, use of uniformly labeled ^{13}C fructose and glucose confirmed the role of each sugar in the formation of pyrazines (Table 7.10). For example, note the dramatic shifts in m/z percentage abundances associated with methylpyrazine. More specifically, increases in m/z can be linked with the incorporation of both two and three ^{13}C carbons into the methylpyrazine molecule. Very similar trends were observed for other pyrazines, and examples for 2,6-dimethylpyrazine and ethylpyrazine are illustrated in Table 7.10. It was observed that fructose and glucose did not perform in the same way regarding the incorporation of labeled carbons.

Pyrazines analysis of natural product suspensions incorporating ^{15}N-labeled amino acids and ^{13}C-labeled sugars revealed the ^{15}N and ^{13}C atoms to be distributed throughout a wide array of volatile pyrazines. The distribution of ^{15}N atoms within pyrazines was unique to the amino acid and the distribution of ^{13}C atoms within pyrazines was linked to the type of sugar employed. Use of either uniformly ^{13}C labeled glucose or uniformly ^{13}C labeled fructose produced similar amounts of incorporation of the ^{13}C label into the pyrazines. In addition to finding ^{13}C atoms in the pyrazines, the label was discovered to be incorporated into sugar thermal degradation products such as 2-butanone; 2,3-butanedione; 2,3-pentanedione; furfural, and 5-methylfurfural. This observation contributed additional data substantiating the possibility that thermal degradation of sugars does occur during the heat treatment of natural product suspensions. Additional experiments with heat-treated tobacco dust suspensions revealed that the anion component of the ammonium salt influenced the yield of pyrazines. For example, ammonium acetate produced approximately three times the amount of pyrazines as did DAP when added to a tobacco suspension followed by heat treatment.

Table 7.10 Percentage Abundances of Selected Pyrazines in Licorice Heat-Treated Formulations Containing Uniformly ^{13}C Labeled Sugars

Methylpyrazine

Ions monitored, m/z	94	96	97	98
Natural formulation	98.72	0.55	0.036	0.37
Uniform ^{13}C glucose	92.32	3.91	3.23	0.54
Uniform ^{13}C fructose	92.09	3.02	4.42	0.46

2,6-Dimethylpyrazine

Ions monitored, m/z	108	109	110	111	114
Natural formulation	82.55	8.87	3.24	2.76	2.58
Uniform ^{13}C glucose	78.61	8.58	3.57	6.74	2.51
Uniform ^{13}C fructose	76.51	8.71	3.74	8.48	2.57

Ethylpyrazine

Ions monitored, m/z	107	108	109	110	111	112	113	114.
Natural formulation	4.78	82.21	7.07	1.40	1.17	1.15	1.12	1.09
Uniform ^{13}C glucose	4.44	78.69	7.04	2.36	4.33	1.15	0.97	1.02
Uniform ^{13}C fructose	4.45	76.77	6.98	2.26	6.13	1.25	1.03	1.19

Summary and conclusions

Reaction conditions for the optimized synthesis of pyrazines in aqueous formulations are described in this chapter. Reaction parameters such as reaction temperature, reaction time, reaction matrix, reaction medium pH, reagent structures, and reagent ratios are described in terms of their impact on product yields, product structures, and product distributions. Employing model compounds containing, for example, uniformly labeled ^{15}N and ^{13}C, led to detailed descriptions of reaction mechanisms. These descriptions help to postulate reaction conditions and reagent ratios necessary to produce aqueous formulations with optimum levels and distributions of pyrazines. Particular attention is given to the type of carbon source (sugars, ketones) employed, as well as the type of nitrogen source (amino acids, ammonium ions, urea) used, to create the pyrazine-rich formulations. The production of pyrazines with both linear and branched alkyl substituents has been directly related to precursor amino acids and selected sugars. New reaction conditions, employing unique carbon sources, for the production of formulations containing unique distributions of pyrazines are described. Further, reaction pathways describing the synthesis of these pyrazines are provided. The emphasis is on not only pyrazine yields but also pyrazine product/structure distributions due in part to the well-documented sensory impact of pyrazines, which is a function of pyrazine structure. Processes and reaction conditions involving both closed pressure vessels and open atmospheric reaction vessels are described with respect to their pyrazine yields. In addition, new production processes for the preparation of aqueous distillates rich in pyrazines are described.

Acknowledgments

The author wishes to thank Dr. Serban Moldoveanu, Dr. Tony Gerardi, Dr. Niraj Kulshreshtha, and Alan Norman for their detailed examination and critique of this manuscript during its preparation.

References

Buchbauer, G., C. T. Klein, B. Wailzer, and P. Wolschann. 2000. Threshold-based structure-activity relationships of pyrazines with bell-pepper flavor. *J Agric Food Chem* 48:4273–4278.

Buttery, R. G., D. J. Stern, and L. C. Ling. 1994. Studies on flavor volatiles of some sweet corn products. *J Agric Food Chem* 42:791–795.

Buckholz Jr., L. L. 1988. The role of maillard technology in flavoring food products. *Cereal Foods World* 33:547–551.

Chen, J., and C.-T. Ho. 1998. Volatile compounds generated in serine-monosaccharide model systems *J Agric Food Chem* 46:1518–1522.

Chen, J., and C.-T. Ho. 1999. Comparison of volatile generation in serine/ threonine/glutamine-ribose/glucose/fructose model systems. *J Agric Food Chem* 47:643–647.

Clarke, R. J. 1986. The flavor of coffee. *Dev Food Sci* 3B(Food Flavors, Pt B) 1.

Coleman III, W. M. 1992. Automated purge-and-trap-gas chromatography analysis of headspace volatiles from natural products. *J Chromatogr Sci* 30:159–163.

Coleman III, W. M., J. L. White, and T. A. Perfetti. 1994. A hyphenated GC-based quantitative analysis of volatile materials from natural products. *J Chromatogr Sci* 32:323–327.

Coleman III, W. M., 1996. A study of the behavior of maillard reaction products analyzed by solid-phase microextraction-gas chromatography-mass selective detection. *J Chromatogr Sci* 34:213–218.

Coleman III, W. M. 1997. A study of the behavior of polar and nonpolar solid-phase microextraction fibers for the extraction of maillard reaction products. *J Chromatogr Sci* 35:245–258.

Coleman III, W. M., and T. A. Perfetti. 1997. The roles of amino acids and sugars in the production of volatile materials in microwave heated tobacco dust suspensions. *Beitr Tabakforsch Int* 17:75–95.

Coleman III, W. M., J. L. White, and T. A. Perfetti. 1996. Investigation of a unique commonality from a wide range of natural materials as viewed from the maillard reaction perspective. *J Sci Food Agric* 70:405–412.

Coleman III, W. M., 1999a. SPME-GC–MS detection analysis of maillard reaction products, In *Applications of Solid Phase Microextraction*, ed. J. Pawliszyn, 43, 585. UK: Royal Society of Chemistry.

Coleman III, W. M. 1999b. Automated solid-phase microextraction/gas chromatography/mass selective detection investigations of carbonyl-amine reaction pathways. *J Chromatogr Sci* 37:323–329.

Coleman III, W. M., and S. N. Lawson. 1999. An automated solid-phase microextraction-gas chromatography-mass selective detection approach for the determination of sugar-amino acid reaction mechanisms. *J Chromatogr Sci* 37:383–387.

Coleman III, W. M., and S. N. Lawson. 2000. Synthesis of materials rich in pyrazines employing no sugar. *J Sci Food Agric* 80:1262–1270.

Coleman III, W. M., and H. L. Chung. 2002a. Pyrolysis GC/MS analysis of amadori compounds derived from selected amino acids and glucose. *J Anal Appl Pyrolysis* 62:215–223.

Coleman III, W. M., and H. L. Chung. 2002b. Pyrolysis GC/MS analysis of amadori compounds derived from selected amino acids and rhamnose and glucose. *J Anal Appl Pyrolysis* 63:349–366.

Coleman III, W. M., and T. J. Steichen. 2006. Sugar and selected amino acid influences on the structure of pyrazines in microwave heat-treated formulations. *J Sci Food Agric* 86:380–391.

Coleman III, W. M., and A. R. Gerardi. 2010. U.S. patent application 20100037903, method of preparing flavor and aromatic compounds.

Cox, R. H., H. J. Grubbs, and S. A. Haut. 1987. U.S. Patent #4638816.

Fors, S. 1983. Sensory properties of volatile maillard reaction products and related compounds. In *The Maillard Reaction in Foods and Nutrition*, eds. G. R. Waller and M. S. Feather, Chapter 12, 185. Washington, DC: ACS.

Hemaini, M. A., C. Cerny, and L. B. Fay. 1995. Mechanisms of formation of alkylpyrazines in the maillard reaction. *J Agric Food Chem* 43:2818–2822.

Hwang, H-I., T. G. Hartman, and C.-T. Ho. 1995. Relative reactivities of amino acids in pyrazine formation. *J Agric Food Chem* 43:179–184.

Tsantili-Kakoulidou, A., and L. B. Kier. 1992. A Quantitative Structure-Activity Relationship (QSAR) study of alkylpyrazine odor modalities. *Pharm Res* 9:1321–1323.

Leahy, M. M. 1985. The effects of pH, types of sugars, amino acids and water activity on the kinetics of alkyl pyrazine formation. PhD Thesis. University of Minnesota, MN, USA.

Lee, P. S. 1995. Modeling the maillard reaction *Chapter 7 Flavor Technology Physical Chemistry, Modification and Process* ACS Symposium #610, ACS, Washington, DC, USA.

Onishi, I., A. Nishi, and T. Kakizawa. 1969. U.S. Patent #3478015.

Parliament, T. H., M. J. Morello, and R. J. McGorrin, eds. 1994. *Thermally Generated Flavors, Maillard, Microwave, and Extrusion Processes.* ACS Symposium #543, ACS, Washington, DC, USA.

Pons, I., C. Garrault, J.-N. Jaubert, J. Morel, and J.-C. Fenyo. 1991. Analysis of aromatic caramel. *Food Chem* 39:311–320.

Shibamoto, T., and R. H. Bernhard. 1977. Investigation of pyrazine formation in glucose ammonia model systems. *Agric Biol Chem* 41: 143–153; *J Agric Food Chem* 25:609–614.

Suwa, K., H. Satoh, and A. Shida. 1973. U.S. Patent #3722516.

Teranishi, R. R. A. Flath, and H. Sugisawa. 1981. *Flavor Research, Recent Advance.* New York, USA: Marcel Dekker.

Vernin, G. 1982. Chemistry of heterocyclic compounds in flavors and aroma. New York, NY: John Wiley & Sons.

Wagner, R., M. Czerny, J. Bieolhradsky, and W. Grosch. 1999. Structure-odour-activity relationships of alkylpyrazines. *Z Lebensm Unters Forsch A* 208:308–316.

Waller, G. R., and M. S. Feather. 1983. *The Maillard Reaction in Foods and Nutrition* ACS Symposium #215. ACS, Washington, DC, USA.

Warfield, A. H., W. D. Galloway, and A. G. Kallianos. 1975. U.S. Patent #3920026.

Whitfield, F. B. 1992. Volatiles from interaction of maillard reactions and lipids. *Crit Rev Food Sci Nutr* 31:1–58.

Wu, D. L., and J. W. Swain. 1981. U.S. Patent #4306577.

Wu, D. L., and J. W. Swaim. 1983. U.S. Patent #4379464.

Yaylyan, V., and A. Keyhani. 1999. Origin of 2,3-Pentanedione and 2,3-Butanedione in D-Glucose/L-Alanine maillard model systems. *J Agric Food Chem* 47: 3280–3284.

chapter eight

Using automated sequential two-dimensional gas chromatography/ mass spectrometry to produce a library of essential oil compounds and track their presence in gin, based on spectral deconvolution software

Albert Robbat, Jr. and Amanda Kowalsick

Contents

Introduction

For tens of centuries, aromatic plants and their oils have been used as medicines and as agents to flavor and aromatize foods and beverages. By the middle of the twentieth century, western medicine lost interest in

183

plant materials. Today, the market for plant-based materials is booming, in part, because of the public's acceptance of essential oils in therapies to improve mood, reduce stress, and relieve pain. The willingness to try these remedies stems from the premise that natural products can treat medical conditions as effectively as prescribed medicines, but with fewer complications and side effects. Essential oils are advanced as antioxidants, anticonvulsant, antianxiety, antidiabetic, anticancer, antibacterial, antiviral agents and as modalities to prevent and treat cardiovascular diseases.[1-7] Foods, vitamins, and essential oils are the cornerstones of Naturopathic Medicine, which purports that many medical conditions are best treated with foods, beverages, and natural supplements we eat, drink, or smell.

Increasing application of essential oils as pesticides, insecticides, and fumigants to reduce offensive odors produced during manufacturing has led some to question whether essential oils should be tested to determine long-term environmental and human health effects before entering the market.[8,9] Earlier studies of efficacy are based on antidotal data, with few controlled studies or clinical trials focused on understanding mechanisms of action. For example, aromatherapy is used to treat attention deficit disorder (ADD) and attention deficit hyperactivity disorder (ADHD) because treatment is thought to be less abusive to the body than available medicines, but levels of success are not well understood. Although highly touted, scientific studies aimed at understanding if/how lavender oil reduces stress and mood swings is open to debate. For example, one report suggests efficacy (relaxation) is the result of expectancy and not the oil[10] and that music is better at relaxing patients in an emergency room waiting area than the scent of lavender oil.[11] In contrast, others found lavender oil reduces opioid treatment of morbidly obese patients undergoing surgery compared to its absence[12] and that bathing infants in the oil reduces stress and crying time and increases sleep time.[13]

Conflicting studies and lack of peer-reviewed clinical trial data leave the medical community confused and conflicted, which leaves the public at risk. Even when clinical trials are conducted, disputes exist as to outcomes as evidenced by the recent exchange between investigators as to why the oils were not analyzed by gas chromatography–mass spectrometry (GC–MS) prior to treatment.[14,15] The assumption being the oils used in the dissenting author's study was somehow different from oils used in studies with positive outcomes. The dissenting authors countered that GC–MS is not common in trials nor can it provide sufficient clarity.[16] Both investigators are correct. GC–MS alone cannot provide the separation power needed to profile complex essential oils nor can it delineate subtle compositional differences from the same or different suppliers. Without this information how can one draw definitive conclusions on efficacy or inferences of potential mechanisms if the sample itself is unknown? How can synergistic effects or the therapeutic value of individual components be assessed?

This level of detail is important to food, flavor, beverage, and personal care companies who spend billions of dollars analyzing competitor products, developing and modifying new ones, evaluating shelf life, and characterizing off-odors and flavors, as well as establishing libraries of key constituents important to their business. To sell foods, beverages, and pharmaceuticals in the United States, companies must comply with the 2002 Bioterrorism Act. The law requires industry to detect adulterants and environmental pollutants and to track and report the genealogy of their products from starting materials.[17] At the moment, the law is unenforceable. No technology exists to accomplish these tasks easily, routinely, and with known data quality.

Since the advent of modern electronics and software, the analytical instrument industry has developed automated sample preparation and analysis instruments, which have dramatically increased sample throughput rates and improved measurement sensitivity, selectivity, precision, accuracy, and data robustness. Instrumentation has gotten smaller, cheaper, and more reliable but conspicuously missing is the same level of advancement in data analysis software. For plant-derived starting materials and end-products, quantitative detection of low concentration analytes requires data analysis software that can "see through" chemical noise, as each component in the sample is either the matrix itself or a potentially harmful, unwanted chemical substance. The purpose of this chapter is to demonstrate how spectral deconvolution can be used to speciate the subtle differences in essential oil content and track key ingredients through the manufacturing process.

Experimental

Samples

Essential oils: Juniper berry oils were refrigerated and analyzed as received. Oils designated with the same numerical value, for example, JB1a and JB1b, were obtained from the same supplier but different batch lots. To assess component profiles after different exposures, 20 mL aliquots of JB1a were (1) UV radiated at 20°C for 4 h (JB4), (2) allowed to evaporate at 20°C for 4 h (JB5), and (3) held at 60°C for 4 h in a closed container (JB6). Oil samples were analyzed by GC–GC/MS (library building) or GC–MS (quantitative analysis).

Distillates: A handful of juniper berries and an aliquot of JB1b were added to ethanol and tested by a sensory expert before distillation after stirring for 2 h. Two distilled samples were prepared: (1) by adding 47 μL of the JB1b oil, 493 mL neutral grain spirit, and 267 mL water and (2) by adding 18.4 g berries soaked overnight in 493 mL neutral grain spirit and 267 mL water. The final distillate contained 666 mL of juniper berry extract. Samples were distilled in the laboratory under conditions that simulated gin production, neither the beginning nor tail fractions were collected for analysis. All gin samples were analyzed by GC–MS.

Gins: Gins from four different batch lots produced over a 2-year period were obtained from the same manufacturer. Trans-cadina-1, 4-diene, a component in the distillate, was used as the reference compound to determine manufacturing precision. Gins from four different manufacturers were purchased and analyzed to assess the possibility of differentiating one manufacturer's gin from another based solely on juniper berry content. In these experiments, phenanthrene-d10 served as the internal standard (Mix #31006 from Restek Corporation, Bellefonte, PA, USA). All gin samples were analyzed by GC–MS.

Multidimensional gas chromatography/mass spectrometry and the library-building process

A retention time, mass spectrometry library was produced from the compounds found in the juniper berry oils by automated sequential, multidimensional GC–GC/MS. Agilent's GC–MS (Little Falls, DE, USA) model 6890N/5975C was used to analyze the oils. Gerstel's (Mülheim an der Ruhr, Germany) MPS 2 was used to automate the sample injection process. A 2 μL oil injection was used in splitless mode. The GC door was modified to house two low thermal mass column heaters. The GC columns were wrapped with heater and sensor wires and independently heated by separate control modules. The GC columns (Restek Corporation, Bellefonte, PA, USA) were connected to one another by an Agilent Dean's switch, which was housed inside the GC oven at 240°C. The GC heated the transfer lines that connected the injector to the first column, the first column to the Dean's switch, the Dean's switch to the second column, and the flame ionization detector as well as the second column to the MS. A Gerstel cryotrap (−150°C) was used to condense and refocus each heart-cut prior to thermal desorption (ramp 25°C/s to 240°C, hold for 10 min) onto the second column. The following experimental conditions were used.

For column 1, a 30 m × 0.25 mm I.D., 25 μm Rtx-wax film was used. The preheart-cut initial temperature was 60°C (2 min), ramped at 4°C/min to 220°C (10 min). The initial pressure was 31.37 psi (2 min), ramped at 0.45 psi/min to 49.45 psi (10 min). The nominal helium flow rate was 0.7 mL/min. The postheart-cut initial temperature was 220°C, cooled at 80°C/min to 60°C (2 min), ramped at 3°C/min to 220°C (10 min). The initial pressure was 49.45 psi, ramped at 8.92 psi/min to 31.37 psi (2 min), ramped at 0.33 psi/min to 49.45 psi (10 min). The average helium velocity was 29 cm/s.

For column 2, a 30 m × 0.25 mm I.D., 25 μm Rxi-5MS film was used. The preheart-cut initial temperature was 60°C (2 min), ramped at 4°C/min to 220°C (10 min). The initial pressure was 26.10 psi (2 min), ramped at

0.38 psi/min to 41.60 psi (10 min). The nominal helium flow rate was 1.5 mL/min. The postheart-cut initial temperature was 220°C, cooled at 80°C/min to 60°C (2 min), ramped at 3°C/min to 240°C (10 min). The initial pressure was 41.60 psi, ramped at 7.60 psi/min to 26.10 psi (2 min), ramped at 0.29 psi/min to 43.90 psi (10 min). The average helium velocity was 49 cm/s.

For the MS, a 50-minute solvent delay was used so that data acquisition began with sample injection onto the second column. Other operating conditions were the following: heated transfer line 280°C, source 230°C, quadrupole 150°C, scan range 30–350 *m/z* at 8 scans/s, and electron multiplier voltage 1952. The total GC–GC/MS run time was 130 min for each heart-cut, which included the time to rinse the syringe prior to each subsequent injection. The total time of analysis for an oil was 3.5 days.

Tracking juniper berry components in oils and distillates

All samples were analyzed by GC–MS using the same 30 m × 0.25 mm, 25 μm wax or Rxi-5MS columns described above. The same initial temperature for both columns was used, viz., 60°C (2 min). The wax column was temperature programmed at 4°C/min to 220°C (10 min). The ramp for the Rxi-5 column was 3°C/min to 240°C (10 min). The GC was operated in constant flow mode at 1 mL/min, with helium as the carrier gas.

A 1 μL sample injection was used for the oils, based on a 100:1 split ratio. For all laboratory prepared juniper berry distillates, 2 mL of the distillate was diluted with 8-mL deionized water prior to analysis. Commercially prepared gins were analyzed as obtained. A 0.5-mm-thick × 10-mm-long polydimethylsiloxane (PDMS) coated stir bar was used to extract organics from all aqueous/ethanol mixtures after stirring for 2 h at room temperature. The autosampler, operated in splitless mode, was equipped with a programmable temperature vapor inlet (CIS 4) and thermal desorption unit (TDU). The system used a solvent vent for 0.5 min at 50 °C followed by ramping the TDU to 250°C (3 min) at 200°C/min. The CIS was held at −150°C for 0.1 minute, then heated at 12°C/s to 250°C (3 min).

Deconvolution software

The spectral deconvolution algorithms were developed by Robbat at Tufts University.[18–20] New mathematical algorithms were tested, which included compound-specific rather than global filters such as a user-defined minimum peak scan number, user-defined ion background subtraction routines, and hierarchical compound identity classification criteria. For the GC–GC/MS work, every peak in every heart-cut was visually inspected to determine if the mass spectrum was unique and invariant. When this occurred, the peak retention time, 1-minute retention window, and mass spectral pattern for at least four ions were selected through the software for input into the library. When mass spectral patterns overlapped, the deconvolution software was used to obtain clean spectra for the library. National Institute of Standards and Technology (NIST) and Adams libraries were linked to the software to automate the compound identification process. The automated mass spectral deconvolution and identification system (AMDIS) spectral deconvolution software made by NIST was also evaluated.

Results and discussion

Oil Results: Since essential oils are sourced globally, our first objective was to assess the chemical variability of juniper berry oils. Juniper berry is used to flavor tea, beer, and brandy as well as marinades for meat, poultry, and fish. It is also the most abundant plant material in gin. Establishing the chemical signature of the oil is difficult, as its content is dependent on which of the six edible plant species manufacturers use to make the oil.[21,22] The plants' growing environment, age, size, ripeness, and isolation method also influence organic content in oil.[23,24] Also contributing to the oil's chemical makeup are the other plant materials that are part of the isolation process.[25,26] Wide differences in chemical content can lead to off-odors and flavors in food and beverage products and would help explain outcome contradictions in the medical literature, especially as they relate to aromatherapies.

Figure 8.1 shows the total ion current (TIC) chromatograms for seven juniper berry oils obtained from three suppliers (1–3) and for different batch lots from the same supplier (1a and 1b) and (3a–3d). Analysts typically look at chromatograms and focus on the 10+ largest peaks to assess similarity and product quality. Oils for JB1a, JB1b, JB2, JB3a and JB3d would be expected to contain the same chemical entities. In fact, there is very little difference among the 42 peaks in each chromatogram. In contrast, the chromatograms for JB3b (bottom trace) and JB3c (top trace) are visually similar but different from the others. Sensory analysis indicated that JB3c oxidized slightly and was the only oil unsuitable to make gin. This is why food, flavor, and fragrance companies rely on sensory (subjective) analysis over chemical (objective) analysis.

A total of 86 juniper berry compounds are reported in the literature.[23,27–33] Although substantial, it is not exhaustive. The following equation describes the relationship between resolution, R, and column efficiency, N, where N is the number of theoretical plates in the column and α and k are the selectivity and retention factors.

$$R = \frac{\sqrt{N}}{4}\left(\frac{\alpha-1}{\alpha}\right)\left(\frac{k}{k+1}\right)$$

If column length doubles, resolution increases by $\sqrt{2}$. Although favorable, the attendant increase in separation time could lead to the loss of low concentration analytes if band broadening occurs. This is a problem

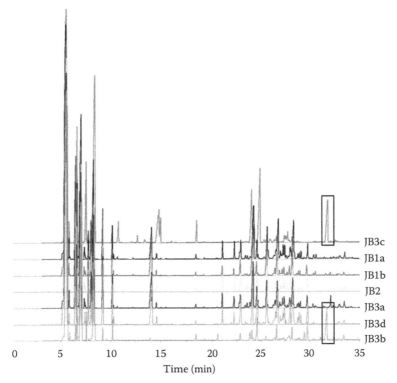

Figure 8.1 Total ion current (TIC) chromatograms of juniper berry oils from the same and different suppliers.

for trace-level, low sensory threshold organics who contribute flavor and aroma. Another approach is to increase selectivity by using dissimilar stationary phases. By automating the GC–GC/MS process, the autosampler injects sequential sample portions from predetermined first column time intervals to the second column. In this study, subsequent injections occur after the preceding sample portion (or heart-cut) elutes from the second column. Automated sequential GC–GC/MS offers higher resolution over other separation choices, but it is time consuming, as it is a function of the total number of heart-cuts and the cumulative column runtimes:

$$T(\text{min}) = \sum_{n=1}^{x}[(n + t_{\text{heart-cut}}) + t_1] + t_2$$

where n is the first heart-cut, $t_{\text{heart-cut}}$ is the time period, t_1 and t_2 are the first and second column GC run times, and x is the total number of heart-cuts defined by t_1/t_{cut}. It is not our intention to compare different multidimensional techniques in this chapter; for that refer to historical[34,35] and updated reviews.[36–44] Nonetheless, theory suggests that sample injected onto stationary phases with different separation mechanisms offers improved opportunities to resolve coeluting compounds and to make matrix-specific libraries.

To detect as many compounds as possible, we purposely overloaded the wax column by splitless injection of 2 μL of JB1a; see flame ionization chromatogram in Figure 8.2. Two examples of TIC chromatograms on Rxi-5 are also shown in figure. Notice the wide unresolved wax peaks between 19 min and 27 min compared to the sharp peaks on Rxi-5 for cut 22. Each 1-minute sample portion produced more than 20 peaks. Even wax heart-cut 14, which visually appears uninteresting, produced more than 20 peaks on the second column. For the most part, compounds that coeluted on one phase separated on the other and when compounds coeluted, we used spectral deconvolution to obtain clean spectra.

Every peak was inspected to meet the following criteria: (1) The mass spectral relative abundance (RA) must be constant for at least five consecutive scans, that is, RA ≤ 20%. (2) The scan-to-scan variance must produce a small relative error (RE), that is, RE < 5. RE is the measure of how well each scan's ions compare to all other scans in the peak. The closer the calculated value is to zero, the smaller is the difference among the spectra. (3) Identity filters, Q-value (> 95), Q-ratio (< 20%), and best retention time (< 0.1 min) for isomers must meet pre-established settings. The Q-value is an integer between 1 and 100. It measures the total ratio deviation of the absolute value of the expected minus observed ion ratios

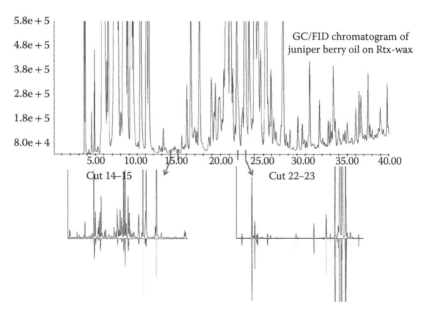

Figure 8.2 Automated sequential GC–GC/MS separation of juniper berry oil (JB1a) and example heart-cuts.

divided by the expected ion ratios times 100 for each ion across the peak. The closer the value is to 100, the greater is the certainty the library and sample spectra match. The Q-ratio is the measure of the molecular (or base) ion to qualifier ion peak area ratios. In this software, confirming ions are normalized to the main ion and are shown as histograms at each scan in the peak. This visual of the deconvolved ion signals provides an easy to interpret analysis of the compound's presence or absence in the sample. Only those scans that fall within the acceptance criteria appear as histograms. When this occurs, the peak area for the main (or quant) ion signal is used to quantify the analyte in the sample.

Figure 8.3 is an expanded view of the peak at 8.66 min for the JB1a oil and illustrates the deconvolution process. When clean fragmentation

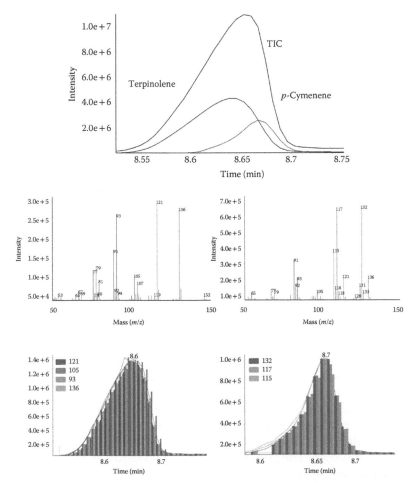

Figure 8.3 The averaged mass spectra and deconvolved reconstructed ion current chromatograms for terpinolene and *p*-cymenene.

patterns from the left-hand side of the peak are averaged, see mass spectrum, and used as target ions, the spectra for all terpinolene scans deconvolve from *p*-cymenene. Similarly, when clean spectra from the right-hand side of the peak are used as target ions, *p*-cymenene spectra are easily deconvolved from terpinolene. The reconstructed ion current (RIC) traces make evident the degree of coelution. Although all of terpinolene's target ions are common to *p*-cymenene and have differing ion ratios, the algorithms correctly deconvolve the spectra and shed ion signal contributions from *p*-cymenene. Identification occurs when the normalized ion ratios at each scan for the quantitative ion (blue) and confirming ions (green, aqua blue, red, and purple) appear at the same height. This occurs when the RE, Q-value, retention time, and Q-ratio meet the identity filters for the compound set by the analyst. The deconvolved peak area provides a more accurate estimate of the analyte concentration, as it no longer includes signal from other compounds in the matrix.

GC–GC/MS analysis of JB1a revealed 190 compounds in the oil; see Table 8.1 for corresponding retention times, retention windows, and mass spectral patterns for the ions used to deconvolve sample components. Although another 20 peaks were detected, peak signals failed to meet the acceptance criteria, which was unfortunate as some of these compounds may be of sensory importance. When the same oils were analyzed by GC/MS, after injecting 200 times less mass on column, a remarkably wide range of juniper berry compounds was found. The *x*-axis in Figure 8.4 shows the number of detected compounds by supplier and batch lot. The *y*-axis depicts the components by retention time and provides an easy to visualize comparison of chemical entities. When oils for JB1a (166 compounds) and JB3c (95 compounds) are compared by component-specific analysis, it is striking how different the oils are despite the remarkable similarities in their chromatograms. Although JB3b was acceptable for manufacturing, its chemical content is conspicuously different from other oils. Chemical speciation of this detail is difficult to obtain by any other means. For example, oils from the same supplier also differ in content, with the range as much as 31 compounds for one of the suppliers. The question of whether JB3c's sensory response is due to missing compounds or contaminants in the oil is outside the scope of this project.

Nonetheless, the technologies used and the results obtained provide the evidence that analysts need to offer sensory experts information as to why oils and ultimately end-products differ chemically from expected outcomes. For example, JB4, JB5, and JB6 were prepared from JB1a and purposely altered to mimic manufacturing conditions that could affect sample composition. The plot shows losses of upwards of 100 compounds when the samples were exposed to UV radiation, elevated temperature, or left to evaporate. Poor oil manufacturing, storage, and handling practices could easily lead to large chemical differences in the final oil. The ability

Table 8.1 Juniper Berry Compounds Found in Oil by GC–GC/MS, with Retention Times and Target Ions

Rxi-5 RT (min)	Compound	Main ion	Confirming ions			
			1	2	3	4
0.951	*n*-Hexane[a]	57	56 (86)	41 (57)	43 (20)	
1.046	Methylcyclo-pentane[a]	56	69 (65)	84 (18)	41 (32)	
1.644	Methylbenzene[a]	91	92 (59)	65 (8)		
1.880	Ethyl butyrate[a]	71	88 (100)	60 (20)	73 (18)	
2.013	2,3-Butanediol[a]	45	75 (17)	43 (8)		
2.234	Furfural[b]	96	95 (98)	39 (23)		
2.286	3,3,5-Trimethyl-Cyclohexene[b]	109	67 (26)	124 (24)	81 (9)	
2.750	*p*-Xylene[b]	91	106 (53)	104 (29)		
2.750	*o*-Xylene[b]	91	106 (51)	105 (22)		
2.839	Isopentyl acetate[b]	70	55 (68)	87 (21)	61 (20)	
3.090	Styrene[b]	104	103 (46)	78 (39)	77 (19)	
3.680	Tricyclene[b,c]	93	121 (49)	136 (28)	105 (22)	79 (19)
3.798	α-Thujene[b,c]	93	92 (36)	77 (33)	136 (15)	
4.299	α-Pinene[b,c]	121	105 (91)	94 (66)	136 (64)	53 (53)
4.447	Camphene[b]	93	121 (73)	107 (29)	67 (25)	
4.521	Thuja-2,4(10)-diene[b]	91	92 (47)	119 (22)	77 (16)	
4.757	Verbenene[b]	91	119 (59)	77 (38)	105 (35)	134 (24)
5.133	Sabinene[b,c]	93	91 (44)	77 (33)	136 (19)	94 (13)
5.207	β-Pinene[b,c]	93	91 (27)	79 (22)	77 (20)	
5.273	1[b,c]	93	79 (65)	121 (53)	107 (45)	67 (25)
5.354	6-Methyl-5-Hepten-2-one[b]	108	69 (57)	111 (41)	55 (35)	
5.738	Myrcene[b,c]	93	69 (64)	41 (61)	67 (12)	53 (11)
5.753	*cis*-2,6-Dimethyl-2,6-octadiene[b]	69	95 (33)	67 (14)	123 (11)	
5.812	Ethyl hexanoate[b]	88	99 (61)	101 (31)	73 (28)	
5.907	Pseudolimonene[b,c]	93	79 (36)	136 (22)	77 (16)	67 (14)
6.040	α-Phellandrene[b,c]	93	91 (59)	77 (40)	92 (35)	136 (26)
6.040	δ-3-Carene[b,c]	93	77 (28)	121 (24)	136 (22)	
6.099	2[b]	117	132 (87)	115 (70)		
6.180	Isoeucolyptol[b]	111	125 (58)	71 (36)	154 (45)	55 (12)
6.210	α-Terpinene[b]	121	93 (85)	136 (52)	91 (45)	
6.549	*p*-Cymene[b,c]	119	134 (29)	117 (14)	115 (11)	

(Continued)

Table 8.1 Juniper Berry Compounds Found in Oil by GC–GC/MS, with Retention Times and Target Ions (*Continued*)

Rxi-5 RT (min)	Compound	Main ion	Confirming ions			
			1	2	3	4
6.844	β-Phellandrene[b]	93	77 (34)	136 (26)	79 (24)	121 (10)
6.844	Limonene[b,c]	93	68 (99)	107 (28)	94 (35)	
6.859	Eucalyptol[b]	111	139 (83)	71 (72)	84 (59)	
6.962	o-Cymene[b,c]	119	134 (31)	91 (23)	117 (13)	
6.962	(Z)-β-Ocimene[b]	93	80 (36)	92 (13)	121 (23)	136 (25)
7.250	(E)-β-Ocimene[b]	93	79 (30)	77 (27)	92 (25)	80 (16)
7.265	3[c]	93	79 (45)	107 (11)	121 (22)	
7.619	γ-Terpinene[b,c]	93	91 (57)	121 (39)	77 (29)	136 (52)
7.744	Ethyl levulinate[b]	99	129 (46)	101 (35)	74 (22)	144 (9)
8.024	4[b]	79	137 (61)	94 (50)	152 (45)	
8.556	Fenchone[b]	81	69 (35)	152 (25)		
8.637	Terpinolene[b,c]	121	136 (97)	93 (85)	105 (26)	
8.666	p-Cymenene[b,c]	132	117 (93)	115 (54)		
8.851	5[b]	79	59 (41)	67 (35)	85 (19)	
8.932	α-Pinene oxide[b]	67	109 (95)	137 (65)	83 (47)	95 (41)
9.116	Ethyl heptanoate[b]	88	113 (55)	101 (40)	73 (20)	
9.116	Perillene[b]	69	150 (80)	81 (80)	53 (27)	
9.131	Linalool[b]	93	55 (45)	121 (35)	80 (31)	
9.264	cis-Thujone[b]	81	110 (83)	55 (35)	152 (13)	
9.271	6[b]	91	92 (40)	65 (16)	63 (10)	
9.492	cis-Rose oxide[b]	139	69 (69)	83 (54)		
9.515	endo-Fenchol/ exo-Fenchol[b]	81	80 (62)	93 (21)	111 (19)	
9.625	trans-Thujone[b]	110	95 (56)	81 (24)	67 (23)	
9.824	trans-Rose oxide[b]	139	69 (41)	55 (32)	83 (27)	
9.900	α-Campholenal[b]	108	93 (75)	95 (31)	67 (24)	
10.083	7[a]	98	111 (45)	55 (45)	83 (40)	84 (30)
10.459	8[b]	91	119 (98)	134 (92)	105 (36)	
10.473	trans-Pinocarveol[b]	92	91 (89)	55 (65)	83 (52)	
10.473	9[b]	91	119 (59)	134 (42)	105 (23)	117 (6)
10.591	10[a]	91	92 (94)	134 (24)		
10.599	1-Terpineol[a]	81	121 (54)	93 (50)	107 (30)	
10.643	Camphor[b,c]	95	81 (68)	108 (46)	152 (33)	
10.820	11[b]	79	110 (77)	95 (53)	109 (43)	
10.940	Camphene hydrate[b]	71	96 (44)	86 (42)	121 (21)	

Table 8.1 Juniper Berry Compounds Found in Oil by GC–GC/MS, with Retention Times and Target Ions (*Continued*)

Rxi-5 RT (min)	Compound	Main ion	Confirming ions			
			1	2	3	4
11.668	Borneol[b]	95	110 (21)	67 (11)	139 (9)	
11.870	(3Z,5E)-1,3,5-Undecatriene[b]	79	80 (66)	150 (66)	78 (24)	
12.347	Terpinen-4-ol[b,c]	71	111 (92)	154 (36)	136 (29)	86 (28)
12.443	12[b]	119	134 (31)	167 (17)		
	Verbenyl ethyl ether[d]	100	119 (79)	137 (50)		
12.576	*p*-Cymen-8-ol[b]	135	43 (19)	91 (16)	150 (13)	
12.782	α-Terpineol[b]	121	136 (88)	59 (55)	81 (46)	
12.886	Myrtenal[b]	79	107 (72)	121 (37)	135 (32)	
12.974	Myrtenol[b]	79	91 (76)	108 (51)	119 (34)	
13.158	Ethyl octanoate[b,c]	88	101 (50)	127 (42)		
13.291	13[b]	95	93 (48)	121 (29)		
13.358	14[b]	112	97 (71)	167 (29)		
13.446	Verbenone[b]	107	135 (84)	91 (63)	150 (42)	
13.800	15[b,c]	93	139 (100)	86 (91)	111 (39)	
13.926	endo-Fenchyl acetate[b,c]	81	121 (51)	80 (39)	107 (20)	
14.471	Citronellol[b]	95	81 (93)	67 (91)	55 (70)	123 (49)
14.582	16[b,c]	112	97 (77)	83 (62)		
14.626	17[b]	119	137 (89)	117 (30)	152 (27)	
14.663	18[a]	137	163 (33)	152 (37)		
14.663	19[b]	163	133 (68)	105 (61)	135 (48)	
14.914	Carvone[b,c]	82	108 (45)	107 (40)	54 (27)	
15.017	Carvacrol, methyl ether[b,c]	149	164 (32)	150 (13)	119 (12)	
15.054	Hexyl isovalerate[b]	85	103 (98)	84 (54)	56 (47)	
15.312	20[a]	109	81 (40)	127 (39)	55 (19)	
15.644	21[b,c]	93	121 (37)	80 (27)	136 (15)	
15.880	Methyl citronellate[b,c]	69	110 (76)	95 (75)	82 (36)	
16.109	*trans*-Ascaridol glycol[b]	109	127 (45)	81 (20)	95 (17)	
16.854	Bornyl acetate[b,c]	95	136 (60)	121 (55)	108 (25)	154 (15)
16.950	22[b]	109	127 (41)	81 (34)		

(*Continued*)

Table 8.1 Juniper Berry Compounds Found in Oil by GC–GC/MS, with Retention Times and Target Ions (*Continued*)

Rxi-5 RT (min)	Compound	Main ion	Confirming ions			
			1	2	3	4
17.068	Carvacrol[b]	135	150 (52)	79 (18)		
17.333	2-Undecanone[b,c]	58	71 (49)	59 (24)	85 (16)	
17.503	Terpinen-4-ol acetate[b]	121	93 (100)	136 (100)		
17.831	Tridecane[b]	57	71 (74)	86 (42)	121 (21)	
18.425	23[b]	112	97 (86)	83 (47)		
18.600	Myrtenyl acetate[b]	91	119 (40)	92 (30)	134 (12)	
18.683	24[a]	140	97 (71)	69 (59)	111 (45)	
19.082	δ-Elemene[b,c]	121	93 (68)	136 (56)	161 (36)	
19.657	α-Cubebene[b,c]	161	105 (88)	119 (84)	204 (25)	
20.048	Citronellyl acetate[b]	95	81 (81)	123 (78)	138 (48)	
20.498	α-Ylangene[b,c]	105	120 (76)	93 (59)		
20.771	α-Copaene[b,c]	161	119 (79)	105 (72)	204 (20)	
	Geranyl acetate[d]	69	68 (35)	121 (34)	136 (20)	
21.177	25[b,c]	93	161 (71)	81 (56)	189 (43)	
21.405	β-Cubebene[b,c]	161	105 (38)	91 (33)	119 (26)	
21.575	β-Elemene[b,c]	93	67 (67)	147 (55)	119 (37)	
21.892	Longifolene[b,c]	161	105 (62)	189 (55)	133 (46)	
22.062	Ethyl decanoate[b,c]	88	101 (60)	157 (39)	155 (36)	
22.202	α-Cedrene[b]	119	204 (36)	161 (16)	105 (16)	93 (16)
22.792	β-Caryophyllene[b,c]	133	93 (99)	91 (91)	120 (51)	
23.065	β-Copaene[b,c]	161	119 (83)	105 (70)	91 (48)	133 (41)
23.345	γ-Elemene[b,c]	121	93 (60)	107 (43)	105 (37)	
23.530	cis-Thujopsene[b]	119	105 (62)	123 (54)	133 (39)	
24.171	α-Humulene[b,c]	93	121 (38)	80 (27)	107 (18)	
24.363	26[b,c]	91	107 (92)	148 (66)	189 (50)	204 (43)
24.363	27[b,c]	161	105 (73)	147 (56)	204 (52)	189 (35)
24.584	(E)-β-Farnesene[b,c]	69	93 (71)	133 (37)		
25.285	Germacrene D[b,c]	161	105 (59)	119 (37)	79 (31)	81 (29)
25.285	28[b,c]	161	105 (53)	204 (38)	133 (16)	
25.536	β-Selinene[c]	105	107 (78)	189 (68)		
25.698	trans-Muurola-4(14),5-diene[b,c]	161	105 (32)	91 (25)	204 (25)	133 (14)
25.927	γ-Muurolene[b,c]	161	105 (84)	119 (57)	204 (44)	
25.927	29[b,c]	161	105 (69)	119 (56)	91 (54)	

Table 8.1 Juniper Berry Compounds Found in Oil by GC–GC/MS, with Retention Times and Target Ions (*Continued*)

Rxi-5 RT (min)	Compound	Main ion	Confirming ions			
			1	2	3	4
26.237	α-Muurolene[b,c]	105	161 (63)	204 (39)	93 (34)	
26.296	Cuparene[b,c]	132	131 (42)	145 (35)	202 (27)	
26.775	γ-Cadinene[b,c]	161	105 (35)	119 (31)	204 (28)	133 (22)
27.115	*trans*-Calamenene[b,c]	159	160 (15)	202 (14)		
27.328	δ-Cadinene[b,c]	161	119 (58)	204 (58)	134 (50)	
27.520	*trans*-Cadina-1,4-diene[b,c]	119	105 (64)	161 (50)	121 (20)	
27.579	30[b,c]	161	133 (58)	135 (37)	147 (20)	
27.727	α-Cadinene[b,c]	105	91 (38)	119 (35)	133 (25)	
27.808	Selina-3,7(11)-diene[b,c]	161	204 (60)	133 (50)	107 (48)	
27.882	α-Calacorene[b,c]	157	142 (49)	141 (29)	200 (20)	156 (17)
28.236	Elemol	93	121 (56)	107 (51)	59 (35)	
28.258	31[b]	79	96 (96)	138 (61)	109 (43)	
28.280	32[b]	79	96 (93)	109 (57)	123 (46)	
28.457	Germacrene B[b,c]	121	93 (66)	105 (59)	161 (39)	
28.671	β-Calacorene[b,c]	157	142 (42)	141 (26)	200 (21)	158 (15)
28.929	Caryophyllenyl alcohol[b]	111	123 (35)	161 (34)		
29.018	33[b,c]	93	69 (66)	107 (65)		
29.409	34[b]	159	187 (64)	145 (55)	202 (35)	
29.534	Caryophyllene oxide[b,c]	79	93 (92)	91 (82)	107 (61)	
29.659	Gleenol[b]	121	222 (46)	108 (39)		
29.859	Salvial-4(14)-en-1-one[b]	128	81 (52)	91 (25)	159 (21)	
30.065	35[b]	93	121 (43)	107 (24)		
30.235	36[b]	107	122 (94)	105 (79)	189 (75)	147 (50)
30.249	Rosifoliol[b]	149	108 (74)	204 (53)	164 (24)	
30.249	37[b]	93	121 (104)	79 (79)	133 (65)	
30.316	Ethyl dodecanoate[b]	88	101 (53)	157 (24)	183 (32)	
30.508	Humulene epoxide II[b,c]	109	96 (90)	138 (88)	67 (53)	
30.699	38[b]	131	105 (83)	159 (77)	145 (60)	

(*Continued*)

Table 8.1 Juniper Berry Compounds Found in Oil by GC–GC/MS, with Retention Times and Target Ions (*Continued*)

Rxi-5 RT (min)	Compound	Main ion	Confirming ions			
			1	2	3	4
30.781	1,10-di-epi-Cubenol[c]	161	119 (55)	179 (55)	105 (38)	
30.803	Junenol[b]	109	161 (71)	204 (56)	179 (44)	
30.803	39[b,c]	93	123 (29)	163 (18)		
31.083	α-Corocalene[c]	185	200 (66)	143 (28)		
31.238	40[b]	166	81 (50)	95 (37)	123 (31)	189 (21)
31.290	1-epi-Cubenol[b,c]	119	161 (82)	105 (52)	105 (41)	
31.437	γ-Eudesmol[b]	189	161 (67)	204 (65)	133 (49)	
31.592	Tetracyclo[6.3.2.0 (2,5).0(1,8)] tridecan-9-ol, 4,4-dimethyl-[a]	136	131 (16)	91 (13)	109 (10)	
31.865	τ-Muurolol[b,c]	161	95 (56)	105 (52)	121 (47)	
31.880	α-Cadinol[b,c]	204	95 (65)	121 (65)	105 (37)	
32.071	α-Muurolol[b,c]	161	119 (51)	105 (46)	204 (31)	
32.271	41[b]	159	177 (50)	117 (41)	220 (38)	131 (35)
32.381	42[b,c]	121	95 (89)	204 (80)	161 (77)	
32.558	*cis*-Calamenen-10-ol[b]	157	175 (38)	203 (36)	142 (26)	158 (16)
32.883	*trans*-Calamenen-10-ol[b]	157	200 (29)	142 (25)	158 (14)	
32.993	43[b]	93	105 (77)	119 (46)	136 (38)	
33.060	Cadalene[b,c]	183	198 (42)	168 (33)	165 (22)	153 (19)
33.259	44[a]	159	177 (44)	131 (35)	220 (31)	
33.510	Amorpha-4,9-dien-2-ol[b]	159	220 (50)	131 (33)	177 (26)	145 (21)
33.613	E-Asarone[b]	208	193 (41)	165 (30)		
33.834	Eudesm-7(11)-en-4-ol[b]	189	204 (55)	222 (43)		
34.314	45[a]	107	147 (57)	135 (56)	162 (51)	
34.793	46[b]	162	220 (78)	149 (69)	187 (57)	202 (45)
36.564	Amorpha-4,7(11)-diene < 2-α-hydroxy->[b]	220	159 (66)	187 (48)		
37.884	Ethyl tetradecanoate[b]	88	101 (70)	213 (28)		
41.292	47[b]	69	93 (78)	133 (62)	229 (50)	

Table 8.1 Juniper Berry Compounds Found in Oil by GC–GC/MS, with Retention Times and Target Ions (*Continued*)

Rxi-5 RT (min)	Compound	Main ion	Confirming ions			
			1	2	3	4
42.200	Sclarene[b]	257	81 (94)	93 (79)	55 (49)	
43.020	49[b]	135	272 (84)	107 (75)	95 (69)	
43.040	50[a]	175	157 (46)	193 (22)		
43.195	51[b]	69	91 (68)	119 (57)	41 (29)	
44.102	52[c]	79	81 (96)	201 (31)	135 (60)	
44.287	53[b]	69	93 (91)	91 (65)	229 (46)	
44.707	Ethyl hexadecanoate[b]	88	101 (61)	241 (28)		
46.190	Abietatriene[b,c]	255	159 (63)	173 (62)	270 (33)	

[a] Compounds found with five peak scans by GC–GC/MS based on 2 μL injection.
[b] Compounds found by GC–MS based on 1 μL splitless injection.
[c] Distillate compounds from oil and botanical.
[d] Juniper berry compounds enhanced from other essential oils and/or botanicals in gin.

Figure 8.4 GC–MS component analysis of fresh, tainted and adultered juniper berry oils from different suppliers and batch lots.

to identify and quantify low concentration organics in complex matrixes offers the best possibility of correlating quantitative sensory descriptors with total product profiling results.

To confirm these results, JB1a was analyzed by GC–MS using the wax column. Spectral deconvolution of the data without the aid of retention windows to help limit the time-axis window was performed. Although the software was unable to determine the retention order of isomers, all 166 compounds were found. Regression of the wax and Rxi-5 retention times produced a Pearson cross-correlation coefficient, r, of 0.89. No compounds eluted early or late on one column with late or early elution on the other column ($r = -1.0$); 91 compounds produced a highly positive and strongly correlated retention behavior, $r = 0.99$; 62 components yielded a weaker, positive relationship, $r = 0.83$; only 13 analytes when regressed produced moderate elution characteristics, $r = 0.41$. The deconvolution software correctly identified minor components in the presence of high concentration oil compounds. For example, limonene produced a peak signal of 10^7 compared to β-phellandrene, eucalyptol, *o*-cymene, and (Z)-β-ocimene, whose peak signals were all $< 10^4$. See Table 8.1 for retention times.

The same JB1a GC–MS data file was analyzed using the AMDIS spectral deconvolution software. AMDIS extracts mass spectra by fitting a least-square regression model to the ion chromatograms.[45] Based on a match factor of 70, which was the criterion used to compare sample versus library spectra, AMDIS detected 96 compounds. Recall there were 166 analytes identified in the oil. Of the 96 reported by AMDIS, 70 were identified correctly, which means there were 26 false positives and 96 false negatives. When the AMDIS and IST peak areas were compared, 15 compounds had relative percent differences (RPD) of $\leq 15\%$, 13 were between 16% and 30%, with the remaining 42 compounds $> 30\%$. The comparison produced 43 positive RPDs (underestimation) and 27 negative RPDs (overestimation), which suggests AMDIS peak areas are biased toward underestimating analyte concentration in the sample.

Distillate and Gin Results: To determine which juniper berry components survived distillation, oil and berry grain alcohol extracts were prepared, distilled, and analyzed by GC–MS. Figure 8.5 shows the component profiles for the JB1b oil, oil and berry distillate, and their residuals. 108 and 111 compounds were found in the oil and berry chromatograms, respectively; 90 of them were detected in both distillates. There are 18 and 21 unique oil and berry organics, respectively, in the distillate mixtures.

A total of 80 juniper berry compounds were detected in four gins produced over a 2-year period by the same manufacturer; 69 of them can be attributed to oil, berry, or both, as they were common to both distillates. See gin profiles in Figure 8.5. Five compounds were due solely to the oil. Four compounds were due solely to the berries. Verbenyl ethyl ether and geranyl acetate were found in the oil by GC–GC/MS but not by GC–MS due to the higher mass amount injected on column for the former versus the latter. We suspect their presence is due to enrichment from other botanicals and oils in the gin. The question of whether these eleven juniper berry compounds influence gin flavor is outside the scope of this study.

Trans-cadina-1, 4-diene served as the reference compound to determine manufacturing consistency of juniper berry concentration in the four

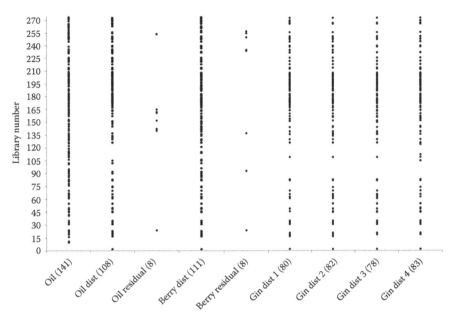

Figure 8.5 GC–MS component analysis of juniper berry oil (JB1b) and its distillate, juniper berry distillate and their residuals, and commercial gins samples.

gins. The peak ratio for each component was computed by dividing the peak area of the compound by the peak area of the reference compound. Based on the 76 analytes detected in all four gins, the peak area ratio relative standard deviation (RSD) was ≤ 50% for 39 compounds, between 51% and 100% for 28, and > 100% for 9 others. High concentration volatiles produced the poorest percent RSDs. Work is in progress to optimize extraction conditions and to assess large volume total desorption dynamic headspace as an alternative sample injection technique, as Twister extraction is selective and its efficiency is compound dependent. Nonetheless, these results suggested a manufacturing precision of 100% which accounted for 90% of the juniper berry content in this manufacturer's gin. We used the manufacturing precision to distinguish one manufacturer's gin from another.

Since gin manufacturers use different essential oils, botanicals, and concentrations, thereof, we added a known concentration of phenanthrene-d_{10} (internal standard) to each manufacturer's gin prior to analysis. The peak area ratio for each compound was calculated by dividing that compound's peak area by the internal standard peak area. Then, we divided the peak area ratio for each compound found in gins 1, 2, and 3 by the peak area ratio for the same compound in gin 4. Deconvolution of GC–MS data of gins from four different manufacturers produced 80 ± 5 compounds. Although the range was somewhat wider than that produced by manufacturer above, this finding was not surprising as the end products are a function of the botanicals and/or oils each used by each manufacturer. The manufacturer precision established above provided the benchmark to determine juniper berry content differences among the manufacturers. For example, if the peak area ratio of gin 1 compounds divided by gin 4 compounds is 1 for all components, the juniper berry content is the same for both gins. If, on the other hand, the ratio is > 2, the juniper berry concentration in gin 1 is materially different from gin 4. This means the concentration of each compound in gin 1 is at least twice that of gin 4.

Gin 1 contains much more juniper berry than gin 4, as the peak area ratios for 49 components in gin 1 are 3.5 times that of gin 4, which is much greater than the manufacturing precision error range. Figure 8.6 shows the juniper berry compounds detected in each gin. The *y*-axis scale makes it seem like there is more juniper berry in gin 1 than gin 4. Component-specific analysis leaves no doubt. Gin 3 also has more juniper berry than gin 4, since 40 compounds have peak area ratios > 2, with 24 of them > 3.5 times. When comparing gins 1 and 3, the former contains more juniper berry than the latter, as 31 of 43 compounds have concentrations > 3.5 times. Similarly, gin 2 contains more juniper berry than gin 4. But, when gins 2 and 3 are compared, there is an equal number of compounds with peak area ratios above 2 and below 0.5, which means both gins contain about the same amount of juniper berry concentration. We suspect organics from other botanicals and oils common to juniper berry may contribute to the small differences in concentration in each gin or that the wide variation in juniper berry content accounts for these differences. Work is in progress to analyze other typical gin additives to evaluate this hypothesis.

Conclusion

Food, flavor, and personal care companies often assess their own products and compare it against competitors based on insufficient data. These companies also rely on sensory rather than chemical analysis to make product decisions. This study demonstrates that automated sequential GC–GC/MS can be used to produce a vast library of compounds from plant-derived materials, which can be tracked from starting material to end-product by GC–MS and spectral deconvolution software. The results demonstrate that the chemical content of juniper berry and, presumably, other essential oils can vary greatly among and from within the same supplier. Such widely differing content will lead to large variability in final products and manufacturing precision as well as outcomes in medical studies. We showed it is possible to identify every detectable component in a complex mixture such as juniper berry and its oil to meet the spirit of the 2002 Bioterrorism Act and requirements in the law. Work is in progress to subtract library mass spectra from TIC signals after detection. Such

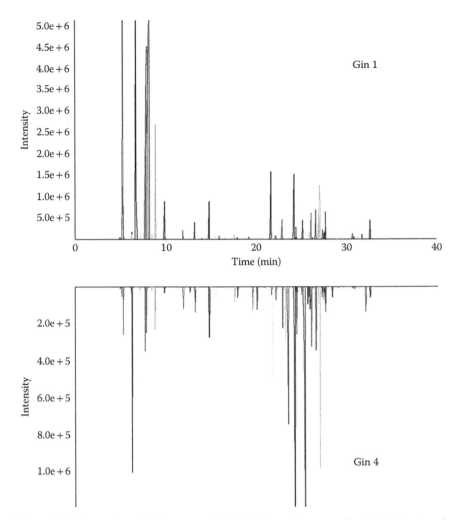

Figure 8.6 Reconstructed ion current (RIC) chromatrograms for Gin 1 (top) and Gin 4 (bottom).

a tool would greatly reduce uncertainties using matrix-specific libraries as well as standard libraries such as NIST, Wiley, and Adams when identifying unknowns and adulterants.

Acknowledgments

We thank Pernod Ricard for providing the financial support, samples, and sensory analysis for this project. University authors are especially appreciative of the many discussions with Pernod Ricard staff, whose contributions helped focus the research objectives described herein. University authors greatly appreciate the contributions made to this research by Gerstel GmbH, Gerstel, Inc., Agilent Technologies, and Restek for their equipment, supplies, and software.

References

1. Edris, A. E. 2007. Pharmaceutical and therapeutic potentials of essential oils and their individual volatile constituents: A review. *Phytother Res* 21:308.
2. Kritsidima, M., T. Newton, and K. Asimakopoulou. 2010. The effects of lavender scent on dental patient anxiety levels: A cluster randomised-controlled trial. *Community Dent Oral Epidemiol* 38:83.
3. Imanishi, J., H. Kuriyama, I. Shigemori, S. Watanabe, Y. Aihara, M. Kita, K. Sawai et al. 2009. Anxiolytic effect of aromatherapy massage in patients with breast cancer. *eCAM* 1:123.
4. Oliveira, J. S., L. A. Porto, C. S. Estevam, R. S. Siqueira, P. B. Alves, E. S. Niculau, A. F. Blank, R. N. Almeida, M. Marchioro, and L. J. Quintans-Junior. 2009. Phytochemical screening and anticonvulsant property of Ocimum basilicum leaf essential oil. *Bol Latinoam Caribe Plant Med Aromat* 8:195.
5. Wahab, A., R. Ul Haq, A. Ahmed, R. Alam Khan, and M. Raza. 2009. Anticonvulsant activities of nutmeg oil of Myristica fragrans. *Phytother Res* 23:153.
6. Ramos Goncalves, J. C., F. de Sousa Oliveira, R. B. Benedito, D. Pergentino de Sousa, R. Nobrega de Almeida, and D. A. Machado de Araujo. 2008. Antinociceptive activity of (−)-carvone: evidence of association with decreased peripheral nerve excitability. *Biol Pharm Bull* 31:1017.
7. Dobetsberger, C. M. 2010. *Effects of Essential Oils on the Central Nervous System—An Update.* Master's Thesis. Vienna: Universitat Wien.
8. Adams, T. B., R. L. Smith, S. M. Cohen, J. Doull, V. J. Feron, J. I. Goodman and L. J. Marnett, et al. 2005. *Food Chem Toxicol* 43:345.
9. Bakkali, F., S. Averbeck, D. Averbeck, and M. Idaomar. 2008. Biological effects of essential oils-A review. *Food Chem Toxicol* 46:446.
10. Howard, S., and B. M. Hughes. 2008. Expectancies, not aroma, explain impact of lavender aromatherapy on psychophysiological indices of relaxation in young healthy women. *Br J Health Psychol* 13:603.
11. Holm, L., and L. Fitzmaurice. 2008. Emergency department waiting room stress: can music or aromatherapy improve anxiety scores? *Pediatr Emerg Care* 24:836.
12. Bekker, A., J. T. Kim, C. J. Ren, G. A. Fielding, A. Pitti, T. Kasumi, M. Wajda and A. Lebovits. 2007. Treatment with lavender aromatherapy in the post-anesthesia care unit reduces opioid requirements of morbidly obese patients undergoing laparoscopic adjustable gastric banding. *Obes Surg* 17:920.
13. Field, T., T. Field, C. Cullen, S. Largie, M. Diego, S. Schanberg and C. Kuhn. 2008. Lavender bath oil reduces stress and crying and enhances steep in very young infants. *Early Hum Dev* 84:399.
14. Kiecolt-Glaser, J. K., J. E. Graham, W. B. Malarkey, K. Porter, S. Lemeshow and R. Glaser. 2008. Olfactory influences on mood and autonomic, endocrine, and immune function. *Psychoneuroendocrinology* 33:328.

15. Hedayat, K. M., and M. Tsifansky. 2008. Letter to the Editor: Olfactory influences on mood and autonomic, endocrine, and immune function. *Psychoneuroendocrinology* 33:1302.
16. Kiecolt-Glaser, J. K., J. E. Graham, W. B. Malarkey et al. 2008. Response to Letter to the Editor regarding "Olfactory influences on mood and autonomic, endocrine, and immune function."*Psychoneuroendocrinology* 33:1303.
17. Title III-Protecting safety and security of food and supply, Subtitle A-Protection of food supply, US Food and Drug Administration: http://www.fda.gov/RegulatoryInformation/Legislation/ucm155769.htm
18. Robbat Jr., A. 2000. Productivity enhancing mass spectral data analysis software for high throughput laboratories: Simultaneous detection of volatile and semivolatile organics by GC/MS. *Environ Test Anal* 9:15.
19. Robbat Jr., A., S. Smarason, and Y. V. Gankin. 1999. Fast gas chromatography/mass spectrometry in support of risk-based decisions. *Field Anal Chem Technol* 3:55.
20. Gankin, Y. V., A. Gorshteyn, S. Smarason, and A. Robbat Jr. 1998. Time-condensed analyses by mass spectrometry. *Anal Chem* 70:1655.
21. Adams, R. 2004. *Junipers of the World: The Genus Juniperus.* Victoria: Trafford.
22. Farjon, A. 1998. *World Checklist and Bibliography of Conifers.* Royal Botanic Gardens, Kew.
23. Rezzi, S., C. Cavaleiro, A. Bighelli, L. Salgueiro, A. Proenca da Cunha, and J. Casanova. 2001. Intraspecific chemical variability of the leaf essential oil of Juniperus phoenicea subsp. turbinata from Corsica. *Biochem Syst Ecol* 29:179.
24. Chatzopoulou, P., and S. Katsiotis. 1995. Procedures influencing the yield and the quality of essential oil from Juniperus communis L. berries. *Pharmacol Acta Helv* 70:247.
25. Gonny, M., C. Cavaleiro, L. Salgueiro, and J. Casanova. 2006. Analysis of Juniperus communis subsp. alpina needle, berry, wood and root oils by combination of GC, GC/MS and ¹³C-NMR. *Flavour Fragance J* 21:99.
26. Angioni, A., A. Barra, M. Russo, V. Coroneo, S. Dessi, and P. Cabras. 2003. Chemical composition of the essential oils of Juniperus from ripe and unripe berries and leaves and their antimicrobial activity. *J Agric Food Chem* 51:3073.
27. Marongiu, B., S. Porcedda, A. Piras, G. Sanna, M. Murreddu, and R. Loddo. 2006. Extraction of Juniperus communis L. ssp. nana Willd. essential oil by supercritical carbon dioxide. *Flavour Fragance J* 21:148.
28. Loizzo, M., R. Tundis, F. Conforti, A. Saab, G. Statti, and F. Menichini. 2007. Comparative chemical composition, antioxidant and hypoglycaemic activities of Juniperus oxycedrus ssp. oxycedrus L. berry and wood oils from Lebanon. *Food Chem* 105:572.
29. Guerra Hernandez, E., M. Del Carmen Lopez Martinez, and R. Garcia Villanova. 1987. Determination by gas chromatography of terpenes in the berries of the species Juniperus oxycedrus L., J. thurifera L. and J. sabina L. *J Chromatogr A* 396:416.
30. Milos, M., and A. Radonic. 2000. Gas chromatography mass spectral analysis of free and glycosidically bound volatile compounds from Juniperus oxycedrus L. growing wild in Croatia. *Food Chem* 68:333.
31. Adams, R. P., J. Altarejos, C. Fernandez, and A. Camacho. 1999. The leaf essential oils and taxonomy of Juniperus oxycedrus L. subsp. oxycedrus, subsp. badia (H. Gay) Debeaux, and subsp. macrocarpa (Sibth. & Sm.) Ball. *Journal of Essential Oil Research* 11:167.

32. Adams, R. P., A. F. Barrero, and A. Lara. 1996. Comparisons of the leaf essential oils of Juniperus phoenicea, J. phoenicea sub-sp. eu-mediterranea Lebr. & Thiv. and J. phoenicea var. turbinata (Guss.) Parl. *Journal of Essential Oil Research* 8:367.

33. Chatzopoulou, P. S., and S. T. Katsiotis. 1993. Study of the essential oil from Juniperus communis "Berries" (Cones) growing wild in Greece. *Planta Med* 59:554.

34. Bertsch, W. 1999. Two-dimensional gas chromatography. Concepts, instrumentation, and applications—Part 1: Fundamentals, conventional two-dimensional gas chromatography, selected applications. *J High Resolut Chromatogr* 22:647.

35. Bertsch, W. 2000. Two-dimensional gas chromatography. Concepts, instrumentation, and applications—Part 2: Comprehensive two-dimensional gas chromatography. *J High Resolut Chromatogr* 23:167.

36. Blumberg, L. M. 2008. Accumulating resampling (modulation) in comprehensive two-dimensional capillary GC (GC×GC). *J Sep Sci* 31:3358.

37. Blumberg, L. M., F. David, M. S. Klee, and P. Sandra. 2008. Comparison of one-dimensional and comprehensive two-dimensional separations by gas chromatography. *J Chromatogr A* 1188:2.

38. Mondello, L., A. C. Lewis, and K. D. Bartle. 2002. *Multidimensional Chromatography*. John Wiley & Sons: West Sussex, England.

39. Marriott, P. J., P. D. Morrison, R. A. Shellie, M. S. Dunn, E. Sari, and D. Ryan. 2003. Multidimensional and comprehensive two-dimensional gas chromatography. *LCGC Eur* 16:23.

40. Adahchour, M., J. Beens, R. J. J. Vreuls, and U. A. T. Brinkman. 2006. Recent developments in comprehensive two-dimensional gas chromatography (GC×GC) Part I: Introduction and instrumental set-up. *TrAC, Trends in Analytical Chemistry* 25:438.

41. Adahchour, M., J. Beens, R. J. J. Vreuls, and U. A. T. Brinkman. 2006. Recent developments in comprehensive two-dimensional gas chromatography (GC×GC). Part II. Modulation and detection. *TrAC, Trends in Analytical Chemistry* 25:540.

42. Adahchour, M., J. Beens, R. J. J. Vreuls, and U. A. T. Brinkman. 2006. Recent developments in comprehensive two-dimensional gas chromatography (GC×GC). Part III. Applications for petrochemicals and organohalogens. *TrAC, Trends in Analytical Chemistry* 25:726.

43. Adahchour, M., J. Beens, R. J. J. Vreuls, and U. A. T. Brinkman. 2006. Recent developments in comprehensive two-dimensional gas chromatography (GC×GC): IV. Further applications, conclusions and perspectives. *TrAC, Trends in Analytical Chemistry* 25:821.

44. Mondello, L., P. Q. Tranchida, P. Dugo, and G. Dugo. 2008. Comprehensive two-dimensional gas chromatography-mass spectrometry: A review. *Mass Spectrom Rev* 27:101.

45. Stein, S. E. 1999. An integrated method for spectrum extraction and compound identification from gas chromatography/mass spectrometry data. *J Am Soc Mass Spectrom* 10:770.

chapter nine

Character-impact flavor and off-flavor compounds in foods

Robert J. McGorrin

Contents

Introduction

Aroma substances that comprise food flavors occur in nature as complex mixtures of volatile compounds. However, a vast majority of volatile chemicals that have been isolated from natural flavor extracts do not elicit aroma contributions that are reminiscent of the flavor substance. For instance, *n*-hexanal is a component of natural apple flavor [1]; however, when smelled in isolation its odor is reminiscent of "green, painty, rancid oil." Similarly, ethyl butyrate provides a nondescript "fruity" aroma to blackberries, raspberries, and pears, but it does not distinctly describe the flavor quality of any of these individual fruits. It has long been the goal of flavor chemists to elucidate the identity of pure aroma chemicals that possess the unique flavor character of the natural fruit, vegetable, meat, cheese, or spice from which they were derived. Frequently, these unique flavor substances are referred to as "character-impact compounds" [2].

A character-impact compound is a unique chemical substance that provides the principal sensory identity of a particular flavor or aroma.

Often, character impact is elicited by a synergistic blend of several aroma chemicals. When tasted or smelled, the character-impact chemical, or group of chemicals, contributes a recognizable sensory impression even at the low concentration levels typically found in natural flavors (e.g., vanillin in vanilla extract and diacetyl in butter) [3,4]. In some instances, flavor concentration and food context are very important. For example, although at high concentrations 4-mercapto-4-methyl-2-pentanone (cat ketone) has an off-odor associated with cat urine, when it is present in reduced levels in the context of a Sauvignon Blanc wine it conveys the typical flavor impression of the Sauvignon grape [5].

For many foods, character-impact flavor compounds are either unknown or have not been reported to date. Examples of such foods include cheddar cheese, milk chocolate, and sweet potatoes. For these foods, the characterizing aroma appears to comprise a relatively complex mixture of flavor compounds rather than one or two aroma chemicals.

The intent of this chapter is to summarize and update a previous review [6] regarding what is generally known about the chemical identities of characterizing aroma chemicals in fruits, vegetables, nuts, herbs and spices, and savory and dairy flavors. A brief compendium of characterizing off-flavors and taints that have been reported in foods is also discussed.

Character-impact flavors in foods

More than 6000 compounds have been identified in the volatile fraction of foods [7]. The total concentration of these naturally occurring components varies from a few parts per million (ppm) to approximately 100 ppm, with the concentration of individual compounds ranging from parts per billion (ppb) to parts per trillion (ppt). A majority of these volatile compounds do not have significant impact on flavor. For example, more than 800 compounds have been identified in the flavor of coffee, but in general only a small proportion of these substances have a significant contribution to its sensory flavor profile [8].

The ultimate goal of flavor research is to identify and classify unique aroma chemicals that contribute to the characteristic odor and flavor of foods. This knowledge enables the flavor industry to better duplicate flavors through nature-identical or biosynthetic pathways and can facilitate better quality control of raw materials by screening appropriate analytical target compounds.

In recent studies, potent aroma compounds have been identified using various gas chromatography–olfactometry (GC–O) techniques, such as combined hedonic aroma response measurements (CHARM) analysis, and aroma extract dilution analysis (AEDA) [9,10]. The flavor compounds identified by these methods are significant contributors to the sensory profile.

In some cases, these sensory-directed analytical techniques have enabled the discovery of new character-impact compounds. However, in other instances key aroma chemicals have been identified that, while potent and significant to flavor, do not impart character impact. For example, in dairy products, chocolate, and kiwifruit, these flavor types appear to be produced by a complex blend of noncharacterizing key aroma compounds.

Knowledge of character-impact compounds enables flavor chemists to use these materials as basic "keys" to formulate enhanced versions of existing flavors. As analytical techniques improve in sensitivity, flavor researchers continue their quest to discover new character-impact flavors that will enable them to develop the next generation of improved flavor systems.

Herb, spice, and seasoning flavors

The original identifications of character aroma compounds were from isolates of spice oils and herbs. Many of these early discoveries paralleled developments in synthetic organic chemistry [11]. The first identifications and syntheses of character flavor molecules include benzaldehyde (cherry), vanillin (vanilla), methyl salicylate (wintergreen), and cinnamaldehyde (cinnamon). A listing of the character-impact compounds found in herb and spice flavors is presented in Table 9.1.

Table 9.1 Character-Impact Compounds in Herbs, Spices, and Seasonings

Character-impact compounds	Chemical Abstracts Services (CAS) Registry number	Occurrence	Reference
trans-p-Anethole	[4180-23-8]	Anise	4
Methyl chavicol (estragole)	[140-67-0]	Basil	12
(S)-(+)-Carvone	[2244-16-8]	Caraway	14
trans-Cinnamaldehyde	[104-55-2]	Cinnamon	4
Eugenol	[97-53-0]	Clove	4
Eugenyl acetate	[93-28-7]	Clove	12
D-Linalool	[78-70-6]	Coriander	4
trans-2-Dodecenal	[20407-84-5]	Coriander	18
Cuminaldehyde	[122-03-2]	Cumin	12
p-1,3-Menthadien-7-al	[1197-15-5]	Cumin	12
(S)-α-Phellandrene	[99-83-2]	Fresh dill	12
3,9-Epoxy-*p*-menth-1-ene	[74410-10-9]	Fresh dill	21
1,8-Cineole (eucalyptol)	[470-82-6]	Eucalyptus	14
Sotolon	[28664-35-9]	Fenugreek	15

(Continued)

Table 9.1 Character-Impact Compounds in Herbs,
Spices, and Seasonings (*Continued*)

Character-impact compounds	Chemical Abstracts Services (CAS) Registry number	Occurrence	Reference
Benzenemethanethiol	[100-53-8]	Garden cress	28
Diallyl disulfide	[2179-57-9]	Garlic	22
Diallylthiosulfinate (allicin)	[539-86-6]	Garlic	22
1-Penten-3-one	[1629-58-9]	Horseradish	4
4-Pentenyl isothiocyante	[18060-79-2]	Horseradish	22
O,S-Diethyl thiocarbonate	[3554-12-9]	Indian cress	29
Allyl isothiocyanate	[57-06-7]	Mustard	4
Propyl propanethiosulfonate	[1113-13-9]	Onion, raw	22
3-Mercapto-2-methylpentan-1-ol	[227456-30-6]	Onion, raw	26
Allyl propyl disulfide	[2179-59-1]	Onion, cooked	22
2-(Propyldithio)-3,4-dimethylthiophene	[126876-33-3]	Onion, fried	22
Carvacrol	[499-75-2]	Oregano	12
Rotundone	[18374-76-0]	Pepper, black	20
L-Menthol	[89-78-1]	Peppermint	4
t-4-(Methylthio)-3-butenyl isothiocyanate	[13028-50-7]	Radish	22
Verbenone	[80-57-9]	Rosemary	14
(R)-(−)-Carvone	[6485-40-10]	Spearmint	14
Safranal	[116-26-7]	Saffron	17
ar-Turmerone	[532-65-0]	Turmeric	16
Thymol	[89-83-8]	Thyme	14
Methyl salicylate	[119-36-8]	Wintergreen	4

A major contributor to the flavor of basil is methyl chavicol (estragole), which provides tealike green, hay, and minty notes [12]. In Italian spice dishes, basil is complemented by oregano, for which carvacrol is a character-impact aroma. Thyme comprises the dried leaves of *Thymus vulgaris*, a perennial of the mint family, for which thymol provides a warm, pungent, and sweetly herbal note with contributions from carvacrol [13]. Interestingly, with fennel (*Foeniculum vulgare*) seed and oil, the essential flavor character is not a single primary compound but is provided by a combination of *trans*-anethole (anise, licorice), estragole (basil, anise), and fenchone (mint, camphor) [13]. In a classic example of the effect of chiral isomers on flavor character, (S)-(+)-carvone imparts caraway flavor, whereas (R)-(−)-carvone provides spearmint flavor [14].

Toasted and ground fenugreek seed is an essential ingredient of curry powders. Sotolon (3-hydroxy-4,5-dimethyl-2(5H)furanone) was recently established as a character-impact flavor component in fenugreek on the basis of its "seasoning-like" flavor note [15]. Its aroma characteristic changes from caramel-like at low concentration levels to currylike at high concentrations. The sensory impact of sotolon is attributed to its extremely low detection threshold (0.3 mg/L in water) and the fact that its concentration in fenugreek seeds is typically 3000 times higher than its threshold.

Turmeric is also primarily used as a spice component in curry dishes, and as a coloring agent in dried and frozen foods. The character-impact compound for turmeric is reported as *ar*-turmerone [16]. Saffron, the dried red stigmas of *Crocus sativus* L. flowers, is utilized to impart both color and flavor, which is described as "sweet, spicy, floral, with a fatty herbaceous undertone." Safranal (2,6,6-trimethyl-1,3-cyclohexadiene-1-carboxaldehyde) is generally considered the character-impact compound of saffron; however, a recent investigation has identified two other potent compounds, 4,4,6-trimethyl-2,5-cyclohexadien-1-one and an unknown compound, possessing "saffron, stale, dried-hay" aroma attributes [17]. Representative structures for spice impact compounds are shown in Figure 9.1.

A key component of both chili powder and curry powder, cumin is the dried seed of the herb *Cuminum cyminum*, a member of the parsley family. Cuminaldehyde is the principal contributor of aroma and flavor to the spice, which imparts a strong musty/earthy character with green grassy notes contributed by *p*-1,3- and 1,4-menthadienals. *trans*-2-Dodecenal, possessing a persistent fatty-citrus-herbaceous odor, is a character-impact component of coriander, along with *d*-linalool [18].

Sesquiterpenes are important contributors to natural flavors. Previously, the aroma of black pepper was attributed to sensory interactions of terpenes and other flavor components [19]. Recently, rotundone was identified as the key aroma-impact compound for black pepper with a strong, spicy peppercorn character, and it is also a source of the peppery, spicy aroma in Shiraz wines [20]. Its aroma detection thresholds were measured as 8 ng/L in water and 16 ng/L in red wine. (+)-α-Phellandrene, the main constituent of dill (*Anethum graveolus* L.), greatly contributes to the sensory impression of the dill herb [12]. However, fresh dill character impact appears to be a synergistic relationship contributed primarily by (+)-α-phellandrene with a modifying effect from "dill ether," 3,9-epoxy-*p*-menth-1-ene [21].

The *Allium* genus includes garlic, onion, leek, and chive. All comprise sulfur-containing character-impact compounds. The aroma-impact constituents of garlic are diallyl disulfide and the corresponding thiosulfinate derivative (allicin), which are enzymatically released from a sulfoxide flavor precursor (alliin) during the crushing of garlic cloves [22].

The flavor chemistry of sulfur compounds in onion is quite complex [23,24]. Polysulfides and thiosulfinates that were initially reported to be flavor significant were later demonstrated to be thermal artifacts from gas chromatographic analysis [25]. Character-impact sulfur compounds have been proposed for fresh, boiled, and fried onion. In raw, fresh onion,

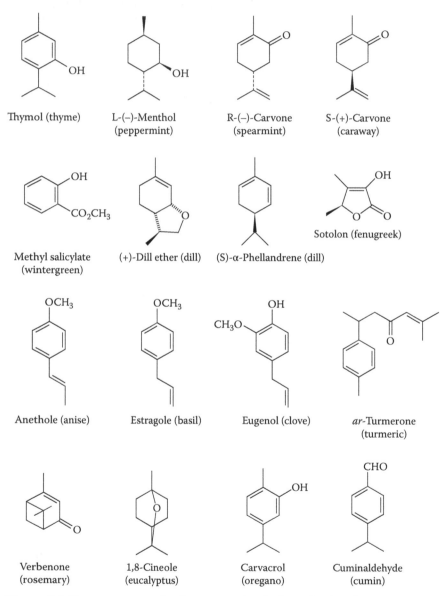

Figure 9.1 Flavor compounds: Representative herb and spice character-impact flavor compounds.

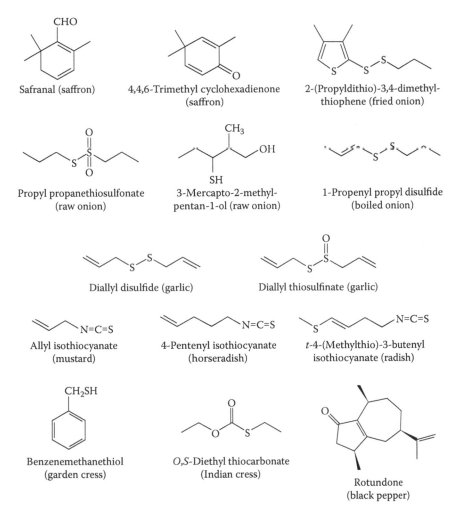

Safranal (saffron)

4,4,6-Trimethyl cyclohexadienone
(saffron)

2-(Propyldithio)-3,4-dimethyl-
thiophene (fried onion)

Propyl propanethiosulfonate
(raw onion)

3-Mercapto-2-methyl-
pentan-1-ol (raw onion)

1-Propenyl propyl disulfide
(boiled onion)

Diallyl disulfide (garlic)

Diallyl thiosulfinate (garlic)

Allyl isothiocyanate
(mustard)

4-Pentenyl isothiocyanate
(horseradish)

t-4-(Methylthio)-3-butenyl
isothiocyanate (radish)

Benzenemethanethiol
(garden cress)

O,S-Diethyl thiocarbonate
(Indian cress)

Rotundone
(black pepper)

Figure 9.1 (*Continued*)

the impact contributors are propyl propanethiosulfinate, propenyl propanethiosulfinate, thiopropanal S-oxide, and propyl methanethiosulfinate
[22,23]. Several compounds contribute to the aroma character of cooked
onion, among which dipropyl disulfide and allyl propyl disulfide show
key impact [22]. Fried onion aroma is formed by heating the latter compound and is characterized by 2-(propyldithio)-3,4-dimethylthiophene,
which has an odor threshold of 10–50 ng/L in water. More recently, a new,
highly potent aroma compound, 3-mercapto-2-methylpentan-1-ol, was
identified in raw onions [26]. The flavor impact of this thiol is strongly
dependent on concentration; at 0.5 ppb it provides a pleasant brothlike,
sweaty, onion, and leeklike flavor, whereas at high levels it provides a

strong, unpleasant onionlike quality. Higher concentrations are formed in cooked onions, and it was also found in other *Allium* species including chives, onions, and leeks but not garlic [27].

Isothiocyanates are character-impact constituents that provide pungency and typical flavor to mustard (allyl isothiocyanate), radish (*trans*-4-(methylthio)-3-butenyl and *trans*-4-(methylthio)butyl isothiocyanate), and horseradish (4-pentenyl and 2-phenylethyl isothiocyanate) [22]. Garden cress (peppergrass) is classified in the *Brassica* (*Cruciferae*/mustard) family, and its leaves and seeds were historically used as salad greens and as a spicy condiment because of their peppery flavor. Benzenemethanethiol was identified as the character-impact compound for the unique flavor of garden cress (*Lepidium sativum*) seed, and it also occurs in the volatile extracts of potatoes [28]. Recently, a novel sulfur volatile, *O,S*-diethyl thiocarbonate, was identified in Indian cress (*Tropaeolaceae* family), providing a "red fruity-sulfury" character [29].

Fruit flavors

The aroma constituents of essential oils from fruits such as lime, lemon, and orange were among the first character-impact compounds identified by flavor chemists. Fruit flavors are a subtle blend of characterizing volatile compounds, supported by fruit sugars, organic acids, and noncharacterizing volatile esters. Fruit aromatics tend to be present in concentrations of greater abundance (< 30 ppm) than other foods, which facilitated early analytical studies. The volatile composition of fruits is extremely complex, and noncharacterizing flavor esters are common across species. A compilation of the character-impact compounds found in fruit flavors is given in Table 9.2.

The combination of ethyl 2-methyl butyrate, β-damascenone, and hexanal is important for the characteristic flavor note of red apples [30,31]. The blend of character-impact flavors combines "apple ester" and "green apple" notes, which fluctuate with apple ripeness and seasonality. β-Damascenone is an unusually potent aroma compound having a threshold of 2 pg/g in water, and it also occurs in natural grape and tomato flavors [31].

Two important character-impact compounds of strawberry flavor are the furanones 2,5-dimethyl-4-hydroxy-2H-furan-3-one (*Furaneol*) and 2,5-dimethyl-4-methoxy-2H-furan-3-one (mesifuran) [32]. However, at various concentrations Furaneol can simulate other flavors, for example, pineapple [14] or Muscadine grape [33] at low levels and caramel at high levels. Mesifuran exhibits a caramel, sherrylike aroma and is a contributor to pineapple, raspberry, and mango. Other important character-impact compounds of strawberry flavor are methyl cinnamate and ethyl 3-methyl-3-phenylglycidate, a synthetic aroma chemical [4,34].

Table 9.2 Character-Impact Flavor Compounds in Fruits

Character-impact compounds	CAS Registry number	Occurrence	Reference
Ethyl-2-methyl butyrate	[7452-79-1]	Apple	4
β-Damascenone	[23696-85-7]	Apple	4, 27
iso-Amyl acetate	[123-92-2]	Banana	4
4-Methoxy-2-methyl-2-butanethiol	[94087-83-9]	Black currant	22
8-Mercapto-p-menthan-3-one	[38462-22-5]	Black currant (synthetic)	47
iso-Butyl 2-butenoate	[589-66-2]	Blueberry	4
Benzaldehyde	[100-52-7]	Cherry	4
p-Tolyl aldehyde	[1334-78-7]	Cherry	4
Ethyl heptanoate	[106-30-9]	Cognac brandy	4
3,5-Dimethyl-1,2,4-trithiolane	[23654-92-4]	Durian	61
Methyl anthranilate	[134-20-3]	Grape, Concord	22
Ethyl-3-mercapto-propionate	[5466-06-8]	Grape, Concord	35
Linalool	[78-70-6]	Grape, Muscat	38
4-Mercapto-4-methyl-2-pentanone	[19872-52-7]	Grape, Sauvignon	5
		Green tea (sencha)	42
		Hops	43
		Grapefruit	44
Nootkatone	[4674-50-4]	Grapefruit	4
1-p-Menthene-8-thiol	[71159-90-5]	Grapefruit	48
Citronellyl acetate	[150-84-5]	Kumquat	52
Citral (neral + geranial)	[5392-40-5]	Lemon, lime	4
α-Terpineol	[98-55-5]	Lime	4
2,6-Dimethyl-5-heptenal	[106-72-9]	Melon	14
Z-6-Nonenal	[2277-19-2]	Melon	30
Methyl N-methyl-anthranilate	[85-91-6]	Orange, mandarin	51
Thymol	[89-83-8]	Orange, mandarin	51
3-Mercapto-1-hexanol	[51755-83-0]	Passion fruit	57, 58
2-Methyl-4-propy-1,3-oxathiane	[67715-80-4]	Passion fruit	22
γ-Undecalactone	[104-67-6]	Peach	4
6-Pentyl-2H-pyran-2-one	[27593-23-3]	Peach	4

(*Continued*)

Table 9.2 Character-Impact Flavor Compounds in Fruits (*Continued*)

Character-impact compounds	CAS Registry number	Occurrence	Reference
Ethyl *trans*-2, *cis*-4-decadienoate	[3025-30-7]	Pear, Bartlett	4
Allyl caproate	[123-68-2]	Pineapple	14
Ethyl 3-(methylthio) propionate	[13327-56-5]	Pineapple	62
Allyl 3-cyclohexylpropionate	[2705-87-5]	Pineapple	14
4-(*p*-Hydroxyphenyl)-2-butanone	[5471-51-2]	Raspberry	4, 53, 54
trans-α-Ionone	[127-41-3]	Raspberry	4, 53, 54
Ethyl 3-methyl-3-phenylglycidate	[77-83-8]	Strawberry	4
Furaneol	[3658-77-3]	Strawberry, Muscat	32, 33
		Guava	60
Mesifuran	[4077-47-8]	Strawberry, mango	32
		Guava	60
(Z,Z)-3,6-Nonadienol	[53046-97-2]	Watermelon	30

Representative chemical structures for fruit flavor-impact compounds are shown in Figure 9.2.

The character-impact component for Concord (*Vitis labrusca*) grape has long been known as methyl anthranilate. Subsequently, ethyl 3-mercaptopropionate was identified in Concord grape, and it possesses a pleasant and fruity fresh grape aroma at low ppm levels [35]. 2-Aminoacetophenone and mesifuran are also significant contributors in Concord to its "foxy" and "candylike" notes, respectively. Ethyl heptanoate elicits an odor reminiscent of cognac brandy [4]. These aroma characters contrast significantly with those resulting from grape varieties (*Vitis vinifera*) used for the production of table wines, whose flavors derive from the specific variety of grape, vinification, maturation, and aging conditions. It is a well-established fact that characteristic wine flavors are produced from secondary metabolites of grapes and yeast during the fermentation process, which increase the chemical and aroma complexity of wines [36]. For example, in Pinot Noir wines, there is no single compound that characterizes the wine's fruity "plum, cherry, ripe blackberry" aroma but rather a complex blend of odorants including 2-phenylethanol, ethyl butyrate, 3-methylbutyl alcohol/acetate, and benzaldehyde, among others [37]. However, other wines made from certain *vinifera* grape varieties possess unique flavor-impact compounds, including "flowery, muscat" character from linalool in Muscat

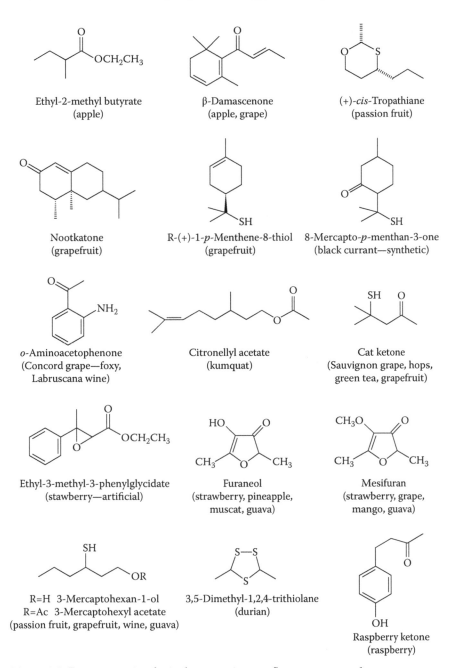

Figure 9.2 Representative fruit character-impact flavor compounds.

wine; "green pepper, earthy" from 3-isopropyl- and 3-isobutyl-2-methoxy pyrazines in Sauvignon Blanc wine; foxy from 2-aminoacetophenone in Labrusca wine; "floral, lychee" from *cis*-rose oxide in Gewurztraminer; and "sulfury" character from 3-mercaptohexan-1-ol, 3-mercaptohexyl acetate, 2-furfurylthiol, and benzyl mercaptan in Grenache, Merlot, and Cabernet wines [38,39, and references cited therein]. Red Tannat wines, produced in southern Uruguay and frequently described as spicy and mintlike, contain 1,8-cineole (eucalyptol) as a flavor character. In Tannat fermentations, 1,8-cineole is produced by chemical rearrangements of limonene or α-terpineol, and levels can be increased by heating to 45°C at wine pH (3.2) [40]. As discussed in the "Introduction" section, 4-mercapto-4-methyl-2-pentanone (cat ketone) provides "catty" character in Sauvignon Blanc (and also Scheurebe) wine at an aroma perception threshold below 3 ppt [5,41]. It is also a characteristic flavorant in Japanese green tea (sencha) [42], a main contributor to the fruity aroma in beer made with certain hops [43], and a significant contributor to the flavor of hand-squeezed grapefruit juice [44].

The mechanism of thioketone formation involves an aldol condensation between two molecules of acetone to yield 4-hydroxy-4-methyl-2-pentanone as shown in Figure 9.3 (**1** in the figure). Dehydration of this intermediate results in 4-methyl-3-penten-2-one (mesityl oxide; **2** in the figure). Hydrogen sulfide (from the degradation of cysteine and methionine amino acids in proteins) can readily undergo a Michael-type addition to the enone moiety of mesityl oxide to generate the thioketone (3 in Figure 9.3) [45].

Black currant flavor is very popular in Europe and is associated with numerous health-related functional foods and alcoholic drinks (cassis liqueur). The key aroma component in black currant is 2-methoxy-4-methyl-4-butanethiol [22]. The "catty/ribes" flavor of black currant (*Ribes nigrum*) was earlier attributed to cat ketone, but it was more recently

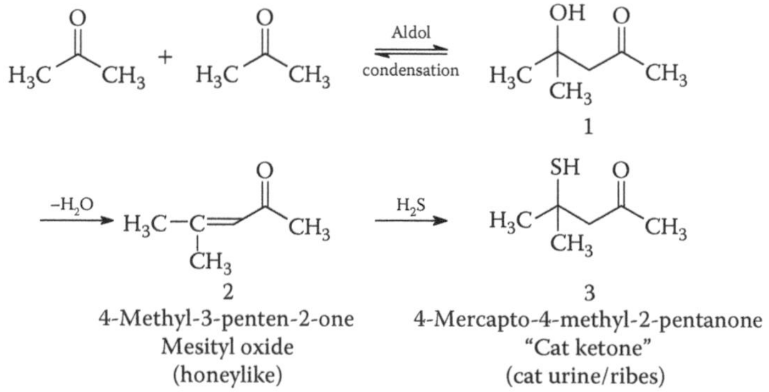

Figure 9.3 Mechanism for the formation of 4-mercapto-4-methyl-2-pentanone (cat ketone).

shown to be absent during flavor and sensory analysis of black currant juice concentrates [46]. The flavor chemical 8-mercapto-*p*-menthan-3-one contributes a powerful black currant, cassis, and cattylike aroma character; however, it has not been identified in the natural fruit [47].

Two character-impact compounds have been proposed for grapefruit flavor, the first being nootkatone, a sesquiterpene. The fresh juicy note of grapefruit juice is attributable to 1-*p*-menthene-8-thiol. This compound has a detection threshold of 10^{-1} ppt, which is among the lowest values reported for aroma chemicals [48]. The (+)-R-isomer was found to have a lower aroma threshold in water than the racemic mixture, and it imparts a pleasant, fresh grapefruit juice character as opposed to the extremely obnoxious sulfur note contributed by the (−)-S-epimer. 3-Mercapto-1-hexanol and its acetate were recently reported to contribute "grapefruit, passion fruit, box tree–like" aromas in fresh grapefruit juice [49].

Among other citrus flavors the basic flavor-impact compound of lemon is citral, a mixture of neral and geranial isomers that together comprise the aroma impression. The flavor character of lime results from a combination of α-terpineol and citral, even though limonene is the most abundant, but sensorially immaterial, volatile in lime and other citrus oils. In contrast to other citrus flavors, orange flavor lacks a specific character-impact compound, and the current belief is that orange flavor is the result of a complex mixture of C_9–C_{13} aldehyde, terpene, and norisoprenoid volatiles blended in specific proportions [50]. For mandarin and tangerine flavors, methyl-*N*-methyl anthranilate and thymol are associated with character impact, with additional contributions from β-pinene and γ-terpinene [51].

Kumquat (*Fortunella japonica* Swingle) is another fruit in the citrus genus that contains characteristic terpene compounds found in other citrus oils. The flesh is quite sour and the peel is often consumed with the flesh. Two sensorially significant diterpene esters, citronellyl formate and citronellyl acetate, were identified in kumquat peel oil by organoleptic evaluation with GC–O. However, citronellyl acetate was found to possess the sensory character that contributes the most to the aroma of kumquat [52].

Lactones have characteristic aromas that are attributable to peach, coconut, and dairy flavors and occur in a wide variety of foods. The γ-lactones, specifically γ-undecalactone and lesser for γ-decalactone, possess intense peachlike odors [14]. A doubly unsaturated δ-decalactone, 6-pentyl-2H-pyran-2-one, also has an intense peach character [4]. As a point of distinction, the C_{10}–C_{12} δ-lactones, particularly the "creamy coconut" note of δ-decalactone, are flavor constituents of coconut as well as cheese and dairy products [14]. Wine lactone (3a,4,5,7a-tetrahydro-3,6-dimethylbenzofuran-2(3H)-one), which conveys an intense "coconutlike, woody, sweet" aroma, has been reported as a background note in Gewurztraminer and Scheurebe white wines [41].

In addition to the character-impact compound of raspberry, 4-(4-hydroxyphenyl)-butan-2-one (raspberry ketone), α- and β-ionone, geraniol, linalool, Furaneol, maple furanone, and other furaneols were concluded to be of importance to raspberry aroma [53,54]. The odor threshold of raspberry ketone was measured at 1–10 µg/kg. The ionones are similar in chemical structure and potencies to β-damascenone [14].

The aroma profiles of Evergreen, Marion, and Chickasaw blackberries are complex, as no single volatile was unanimously described as being characteristically "blackberry" [55,56]. Flavors of importance include ethyl butyrate, Furaneol and other furanones, linalool, and β-damascenone. Blackberry aroma character was shown to differ for the same variety, depending on the growing region [56].

The "tropical" category is one of the most important areas for new discoveries of key flavor impact compounds. Analyses of passion fruit have identified many potent sulfur aroma compounds [22]. Among these is tropathiane, 2-methyl-4-propyl-1,3-oxathiane, which has an odor threshold of 3 ppb [18]. 3-Mercapto-1-hexanol is a powerful odorant reminiscent of citrus and tropical fruit. It was first isolated from passion fruit and contributes to its character impact [57,58]. It has been extensively studied in wine flavor and identified as one of the key aroma compounds in Sauvignon Blanc, Riesling, Gewurztraminer, Cabernet Sauvignon, and Merlot wines [59]. In pink guava fruit, 3-mercapto-1-hexanol (grapefruit aroma) and 3-mercaptylhexyl acetate (black currant aroma) provide tropical character in a nearly racemic ratio, and both the mercaptohexanol enantiomers have extremely low odor thresholds of 70–80 pg/L [60]. These two thiols, in combination with Furaneol, sotolon, mesifuran, and other esters, contribute significantly to the strong and characteristic tropical fragrance of fresh guava fruits. In Indonesian durian fruit, sulfur volatiles dominate the overall aroma perception. From flavor extract dilution analysis, 3,5-dimethyl-1,2,4-trithiolane was identified as the most potent odorant in durian [61]. For pineapple, 2-propenyl hexanoate (allyl caproate) exhibits a typical pineapple character [14]; however, Furaneol, ethyl 3-methylthiopropionate, and ethyl-2-methylbutyrate are important supporting character-impact compounds [62]. The latter ester contributes the background apple note to pineapple flavor. Another character-impact compound, allyl 3-cyclohexylpropionate, has not been discovered in nature, but it provides a sweet, fruity pineapple flavor note [14].

Characterizing flavors for melons include (Z)-6-nonenal, which contributes a typical melon aroma impression, and (Z,Z)-3,6-nonadienol for watermelon rind aroma impact [30]. 2,6-Dimethyl-5-hepten-1-al (Melonal) has not been identified in melon, but it provides a melonlike note in compounded flavors [14]. A more recent study confirmed that the flavor chemistry of muskmelons is more complex, with methyl 2-methylbutyrate and ethyl 2-methylbutyrate providing noncharacterizing "fruity, sweet, and cantaloupe-like" aromas and ethyl 3-(methylthio)propionate contributing "green, fresh melon" notes [63]. Although an important constituent to both varieties, the flavor of sweet cherries (*Prunus aviium*) is less dominated by the character-impact compound benzaldehyde than is the flavor profile of sour cherries (*P. cerasus*) [64].

Vegetable flavors

Recent aroma research has been devoted to the identification of key flavor compounds in vegetables and is the subject of several contemporary reviews [62,65,66]. Cucumbers, sweet corn, and tomatoes are botanically classified as fruits; however, for flavor considerations they are regarded as vegetables, since they are typically consumed with the savory portion of the meal. Overall, the knowledge base of character-impact compounds for vegetables is much smaller than that for other flavor categories and warrants further investigation.

Identifying flavor-impact compounds in vegetables depends considerably on how they are prepared (cutting, blending) and the form in which they are consumed (raw vs. cooked). For example, the character impact of fresh tomato is delineated by 2-*iso*-butylthiazole and (Z)-3-hexenal, with modifying effects from β-ionone and β-damascenone [62]. Alternatively, dimethyl sulfide is a major contributor to the flavor of thermally processed tomato paste [62,65]. Dimethyl sulfide is also a flavor-impact compound for both canned cream and fresh corn, whereas 2-acetyl-1-pyrroline provides a "corn chip" character. Hydrogen sulfide, methanethiol, and ethanethiol may further contribute to the aroma of sweet corn due to their low odor thresholds [66]. A summary of character-impact compounds for vegetable flavors is outlined in Table 9.3.

The character-impact compound of green bell pepper, 2-isobutyl-3-methoxypyrazine, was the first example of a high-impact aroma compound because of its exceptionally low odor threshold of 2 ppt [18]. A compound with a similarly low odor threshold, geosmin is the character impact of red beets and is detectable at a concentration of 100 ppt. Recently, characteristic cooked red bell pepper notes were attributed to (E)-3-heptene-2-thiol (sesame, green, bell peppers, fresh note) and (E)-4-heptene-2-thiol (sesame, coffee, bitterness of peppers) [67]. The flavor of raw peas and peapods is attributed to 2-*iso*-propyl-3-methoxypyrazine [65]. The importance of C_9 aldehydes to the character impact of cucumber flavor was recently confirmed by calculating their odor unit values (ratio of concentration to odor threshold). (E,Z)-2,6-Nonadienal and (Z)-2-nonenal were determined to be the principal odorants of cucumbers [68].

Tomatillo (*Physalis ixocarpa* Brot.) is a solanaceous fruit vegetable similar in appearance to a small green tomato, which is used to prepare green salsas in various Mexican dishes. The character-impact compounds in

Table 9.3 Character-Impact Flavor Compounds in Vegetables

Character-impact compounds	CAS Registry number	Occurrence	Reference
Dimethyl sulfide	[75-18-3]	Asparagus	4
1,2-Dithiacyclopentene	[288-26-6]	Asparagus, heated	65
Geosmin	[19700-21-1]	Red beet	65
4-Methylthiobutyl isothiocyanate	[4430-36-8]	Broccoli	66
Dimethyl sulfide	[75-18-3]	Cabbage	65
S-Methyl methane-thiosulfinate	[13882-12-7]	Sauerkraut	71
2-sec-Butyl-3-methoxypyrazine	[24168-70-5]	Carrot (raw)	65
3-(Methylthio)propyl isothiocyanate	[505-79-3]	Cauliflower	66
3-Butylphthalide	[6066-49-5]	Celery	66
Sedanolide	[6415-59-4]	Celery	66
(Z)-3-Hexenyl pyruvate	[68133-76-6]	Celery	4
Dimethyl sulfide	[75-18-3]	Corn	66
2-Acetyl-2-thiazoline	[29926-41-8]	Corn, fresh	66
(E,Z)-2,6-Nonadienal	[557-48-2]	Cucumber	68
(E)-2-Nonenal	[2463-53-8]	Cucumber	68
2-iso-Butyl-3-methoxypyrazine	[24683-00-9]	Green bell pepper	65
(E)-3-Heptene-2-thiol	[1006684-17-8]	Red bell pepper	67
(E)-4-Heptene-2-thiol	[1006684-18-9]	Red bell pepper	67
1-Octen-3-ol	[3391-86-4]	Mushroom	65
1-Octen-3-one	[4312-99-6]	Mushroom	65
p-Mentha-1,3,8-triene	[18368-95-1]	Parsley	76
2-iso-Propyl-3-methoxypyrazine	[25773-40-4]	Pea (raw)	65
2-iso-Propyl-3-methoxypyrazine	[25773-40-4]	Potato (earthy)	65
3-Methylthiopropanal	[3268-49-3]	Potato (boiled)	21
2-Ethyl-6-vinylpyrazine	[32736-90-6]	Potato (baked)	65
(Z)-3-Hexenal	[6789-80-6]	Tomatillo	69
(E,E)-2,4-Decadienal	[25152-84-5]	Tomatillo	69
2-iso-Butylthiazole	[18640-74-9]	Tomato (fresh)	21
(Z)-3-Hexenal	[6789-80-6]	Tomato (fresh)	62
Bis(methylthio)methane	[1618-26-4]	Truffle	74, 75
1,2,4-Trithiolane	[289-16-7]	Truffle	74

tomatillo were recently established as (Z)-3-hexenal; (E,E)-2,4-decadienal; and nonanal, which impart the fruit vegetable's dominant "green" flavor [69]. Similar to tomato flavor, β-ionone and β-damascenone provide modifying effects in tomatillo; however, tomatillo does not contain 2-*iso*-butylthiazole, a key character-impact compound of tomato.

Among the cruciferous vegetables, cooked cabbage owes its dominant character-impact flavor to dimethyl sulfide. In raw cabbage flavor, allyl isothiocyanate contributes sharp, pungent horseradish-like notes [66]. *S*-Methyl methanethiosulfinate was observed to provide the character impact of sauerkraut flavor, and it occurs in Brussels sprouts and cabbage [70,71]. Compounds likely to be important to the flavor of cooked broccoli include dimethyl sulfide and trisulfide, nonanal, and erucin (4-(methylthio)butyl isothiocyanate) [65]. Cooked cauliflower contains similar flavor components as broccoli, with the exception that 3-(methylthio)propyl isothiocyanate is the characterizing thiocyanate in the former [65]. Key volatile sulfur compounds were identified in the flavor of cooked asparagus [72]. Representative structures for vegetable flavor-impact compounds are presented in Figure 9.4.

Potato flavor is greatly influenced by methods of cooking or preparing the vegetable. Raw potato contains the characteristic "earthy aroma" component, 2-*iso*-propyl-3-methoxypyrazine. A character-impact compound common to boiled and baked potatoes is methional (3-(methylthio)propanal). Baked potatoes contain Maillard products such as 2-ethyl-3-methylpyrazine (earthy, nutty) and 2-ethyl-6-vinylpyrazine (buttery, baked potato) [65]. In potato chips and French-fried potatoes, the potato flavor character of methional is modified by volatile aromatics from frying oils, such as (E,E)-2,4-decadienal, and thermally generated alkyl oxazoles possessing lactonelike flavors [65,73].

Aroma-impact compounds that are universal to all varieties of mushrooms include 1-octen-3-ol (threshold of 1 ppb) and 1-octen-3-one (threshold of 0.05 ppb), both of which have been described as having a fresh, wild-mushroom aroma [18]. However, 1-octen-3-one also possesses a metallic odor, particularly in the context of oxidized oils [65].

The predominant sulfur component in black and white truffle aroma is dimethyl sulfide; however, bis(methylthio)methane and 1,2,4-trithiolane are reported to be unique to white truffle aroma [74]. 1-(Methylthio)propane and 1-methylthio-1-propene were newly identified in black truffle [75].

The characteristic compound for raw carrot is 2-*sec*-butyl-3-methoxypyrazine, which has an extremely low (2-ppt) threshold value. Its sniffing port aroma in GC–O has been described as "raw carroty" [65]. Unsaturated aldehydes contribute to the flavor of cooked carrot, the most significant being (E)-2-nonenal (fatty-waxy) [65].

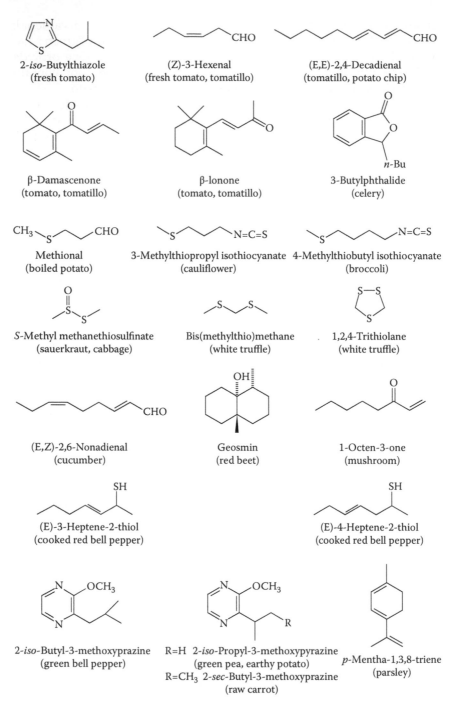

Figure 9.4 Representative vegetable character-impact flavor compounds.

Compounds important to the aroma of celery include two lactones, 3-butylphthalide and sedanolide (3-butyl-3a,4,5,6-tetrahydrophthalide), which provide impact at low concentrations (< 3 ppm). Other terpene hydrocarbons such as β-selinene and limonene are present in greater abundance, but they do not contribute significantly to celery flavor. A recent assessment of its odor threshold by Takeoka [66] suggests that 3-butylphthalide is the most significant character-impact compound for celery.

Key parsley aroma compounds have been reported [76]. The primary flavor contributors were found to include *p*-mentha-1,3,8-triene (terpeny, parsleylike); myrcene (metallic, herbaceous); 2-*sec*-butyl-3-methoxypyrazine (musty, earthy); myristicin (spicy); linalool (coriander); (Z)-6-decenal (green, cucumber); and (Z)-3-hexenal (green).

Maillard-type, brown, and cereal flavors

The flavor characteristics of heated sugar compounds that possess caramel, burnt sugar, and maple notes include a family of structures containing a methyl enol-one group [77]. As discussed in the section *Fruit flavors*, Furaneol (2,5-dimethyl-4-hydroxy-(2H)-furan-3-one) has a sweet, caramel-like, burnt sugar flavor with appreciable fruitiness and occurs in thermally processed foods such as beer, arabica coffee, cocoa, and white bread crust [78]. Maltol (3-hydroxy-2-methyl-4H-pyran-4-one) exhibits a sweet, burnt sugar, caramel note similar to Furaneol, but not as strong or fruity. Ethyl maltol does not occur in nature; however, it possesses a very intense sweet, caramel-like odor, which is four-to-six times more potent than the odor of maltol [14]. Cyclotene (3-methyl-2-cyclopenten-2-ol-1-one) has a strong maple-caramel flavor. 2-Methyltetrahydrofuran-3-one has a very pleasant, sweet-caramel character [18]. Sotolon (4,5-dimethyl-3-hydroxy-2(5H)-furanone) was identified as a potent flavor-impact compound from raw cane sugar. At low concentrations, its aroma character is caramel, maple syrup, and burnt sugar, typical of unrefined cane sugar [79]. At high concentrations, it has the typical seasoning-like aroma of fenugreek or curry. A summary of character-impact compounds for thermally generated flavors is given in Table 9.4.

Characteristic heated flavor compounds arise via the Maillard pathway, the thermally induced reaction between amino acids and reducing sugars. Aroma constituents in chocolate, coffee, toasted cereal grains, wheat bread crust, and popcorn are products of Maillard reactions, in addition to flavors in roasted nuts and meats, which are discussed in "Nut flavors" and "Meat and seafood flavors" sections, respectively. Guaiacols occur as pyrolysis products of carbohydrates or lipids in smoked and charbroiled meats.

Table 9.4 Character-Impact Flavor Compounds in Cooked Flavors and Maillard-Type Systems (Chocolate, Coffee, Caramelized Sugar)

Character-impact compounds	CAS Registry number	Occurrence	Reference
5-Methyl-2-phenyl-2-hexenal	[21834-92-4]	Chocolate	99
2-Methoxy-5-methyl-pyrazine	[2882-22-6]	Chocolate	4, 100
iso-Amyl phenylacetate	[102-19-2]	Chocolate	4, 100
2-Furfurylthiol	[98-02-2]	Coffee	80
		Wine (barrel-aged)	83
Furfuryl methyl disulfide	[57500-00-2]	Coffee (mocha)	18
2-Methyl-3-furanthiol	[28588-74-1]	Wine (barrel-aged)	83
Furaneol	[3658-77-3]	Fruity, burnt sugar	78
Sotolon	[28664-35-9]	Brown sugar	79
Phenylacetaldehyde	[122-78-1]	Honey	100
3-Methylbutanal	[590-86-3]	Malt	88
Maltol	[118-71-8]	Cotton candy	14
Ethyl maltol	[4940-11-8]	Caramel, sweet	14
2-Hydroxy-3-methyl-2-cyclopenten-1-one	[80-71-7]	Maple	14
2-Acetyl-1,4,5,6-tetrahydropyridine	[25343-57-1]	Cracker (saltine)	87
(E,E,Z)-2,4,6-Nonatrienal	[100113-52-8]	Oat flake	95
2-Acetyl-1,4,5,6-tetra-hydropyridine	[25343-57-1]	Popcorn	85, 86
2-Acetyl-1-pyrroline	[85213-22-5]	Popcorn	85, 86
2-Propionyl-1-pyrroline	[133447-37-7]	Popcorn	85
2-Acetylpyrazine	[22047-25-2]	Popcorn	85
2-Acetyl-2-thiazoline	[22926-41-8]	Roasty, popcorn	97
5-Acetyl-2,3-dihydro 1,4-thiazine	[164524-93-0]	Popcorn (model system)	89
3-Thiazolidineethanethiol	[317803-03-5]	Popcorn (model system)	90
2-Acetyl-1-pyrroline	[85213-22-5]	Wheat bread crust	86, 88
		Rice (basmati)	86, 88
2-Ethyl-3,5-dimethyl-5-pyrazine	[13925-07-0]	Potato chip	97
2,3-Diethy-5-methyl-pyrazine	[18138-04-0]	Potato chip	97
2-Vinylpyrazine	[4177-16-6]	Roasted potato	97

Table 9.4 Character-Impact Flavor Compounds in Cooked Flavors and
Maillard-Type Systems (Chocolate, Coffee, Caramelized Sugar) (*Continued*)

Character-impact compounds	CAS Registry number	Occurrence	Reference
(Z)-2-propenyl-3,5-dimethylpyrazine	[55138-74-4]	Roasted potato	97
Guaiacol	[90-05-1]	Smoky	4
4-Vinylguaiacol	[7786-61-0]	Smoky	18

2-Furfurylthiol is the primary character-impact compound for the aroma of roasted arabica coffee [80]. It has a threshold of 5 ppt and smells like freshly brewed coffee at concentrations between 0.01 and 0.5 ppb [81]. At higher concentrations, it exhibits a stale coffee, sulfury note. Other potent odorants in roasted coffee include 5-methylfurfurylthiol (threshold of 0.05 ppb), which smells meaty at 0.5–1.0 ppb and changes character to a sulfury mercaptan note at higher levels [81]. Furfuryl methyl disulfide has a sweet mocha coffee aroma [18]. Whereas other sulfur compounds such as 3-mercapto-3-methylbutyl formate and 3-methyl-2-buten-1-thiol were previously thought to be important factors for coffee aroma, recent studies have confirmed that the flavor profile of brewed coffee is primarily contributed by 2-furfurylthiol, 4-vinylguaiacol, and "malty"-smelling Strecker aldehydes, among others [80,82]. For wines and champagnes aged in toasted oak barrels, 2-furfurylthiol and 2-methyl-3-furanthiol have been identified as providing "roast-coffee" and "cooked meat" aroma characters, respectively [83].

Cereal grains, including wheat, rice, corn, and oats, have characteristic "toasted, nutty" flavors (pyrazines, tetrahydropyridines, pyrrolines, etc.) that are thermally generated through Maillard pathways [84]. The principal impact aroma compounds of freshly prepared popcorn were determined by Schieberle [85] as 2-acetyltetrahydropyridine, 2-acetyl-1-pyrroline, and 2-propionyl-1-pyrroline [86]. The crackerlike aroma of tetrahydropyridine, which exists in two tautomeric forms (2-acetyl-1,4,5,6- and 2-acetyl-3,4,5,6-tetrahydropyridine), was previously identified as the character compound of saltine crackers [87]. The decrease of these compounds during storage was directly correlated with staling flavor. Another "popcornlike" odorant, 2-acetylpyrazine, was determined to have a minor contribution to the aroma of fresh popcorn because of its considerably lower odor activity [88]. Two novel, highly intense "roasty, popcornlike" aroma compounds were recently identified in Maillard model systems: (1) 5-acetyl-2,3-dihydro-1,4-thiazine (odor threshold of 0.06 ppt) from the reaction of ribose with cysteine [89] and (2) N-(2-mercaptoethyl)-1,3-thiazolidine (3-thiazolidineethanethiol) (odor threshold of 0.005 ppt) from the reaction of fructose with cysteamine

[90]. Neither of these characteristic popcorn flavor compounds has been reported in food aromas to date.

The flavor formed by the cooking of fragrant rice (e.g., basmati rice) is described as popcornlike; hence it is not surprising that 2-acetyl-1-pyrroline is the character-impact volatile [88]. In masa corn tortillas, 2-aminoacetophenone provides the character impact resulting from the lime treatment of corn [91], whereas in corn chips its contribution is modified by 2-acetyl-1-pyrroline and unsaturated aldehydes [92].

In addition to Maillard pathways, cereal-like flavor character is provided by trace volatiles formed through lipoxygenase oxidation of unsaturated lipids, such as linoleic and linolenic acids. One of the first studies on oat flavor by Heydanek and McGorrin [93,94] established the importance of C_8–C_{10} unsaturated aldehydes and ketones as key contributors to the flavor of dried oat groats. Recently, Schuh and Schieberle [95] identified the character-impact compound (E,E,Z)-2,4,6-nonatrienal as eliciting the typical "cereal, sweet" aroma of rolled oat flakes. This odorant exhibits an extremely low odor threshold (0.0002 ppt), and the value is among the lowest reported to date. Representative chemical structures for thermally generated flavor-impact compounds are shown in Figure 9.5.

Two compounds that create characteristic odor notes in the pleasant aroma of wheat bread crust have been identified as the popcornlike 2-acetyl-1-pyrroline and 2-acetyltetrahydropyridine [88]. The aroma of the bread crumb portion is principally due to the presence of lipid-derived unsaturated aldehydes such as (E)-2-nonenal and (E,E)-2,4-decadienal, which create stale aromas at high levels. The malty notes that predominate in yeast and sourdough breads are attributed to 2- and 3-methylbutanal and Furaneol [88,96].

"Potato chip" aroma is associated with the pyrazines 2-ethyl-3,5-dimethylpyrazine and 2,3-diethyl-5-methylpyrazine, whereas 2-vinylpyrazine and (Z)-2-propenyl-3,5-dimethylpyrazine provide an intense "roasted potato" odor [97]. A combination of methional, 2-acetyl-1-pyrroline, phenylacetaldeyde, and butanal are important flavor characters for extruded potato snacks [28].

Cocoa and chocolate represent a highly complex flavor system for which no single character impact has been identified. Vanillin and Furaneol contribute to the sweet, caramelized background character of milk chocolate, whereas 3-methylbutanal provides its malty flavor [98]. 5-Methyl-2-phenyl-2-hexenal provides a "deep bitter, cocoa" note and is the aldol reaction product from phenylacetaldehyde and 3-methylbutanal, two Strecker aldehydes formed in chocolate [99]. 2-Methoxy-5-methylpyrazine and *iso*-amyl phenylacetate have "chocolate, cocoa, nutty" and "cocoalike" notes, respectively, and both are used in synthetic chocolate flavors [100]. Systematic studies of key odorants in milk chocolate were performed using AEDA; however, character-impact compounds unique to chocolate flavor were not reported [98,101].

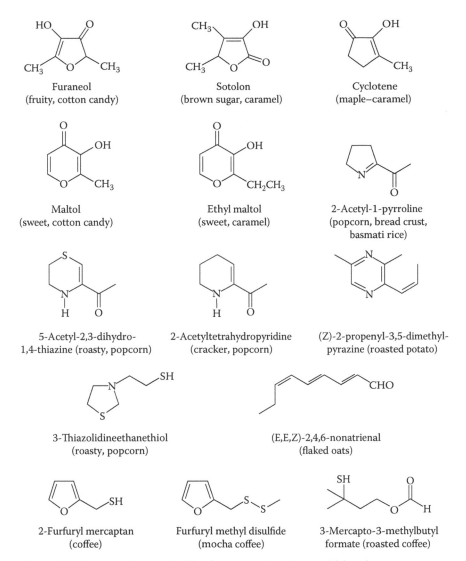

Furaneol
(fruity, cotton candy)

Sotolon
(brown sugar, caramel)

Cyclotene
(maple–caramel)

Maltol
(sweet, cotton candy)

Ethyl maltol
(sweet, caramel)

2-Acetyl-1-pyrroline
(popcorn, bread crust,
basmati rice)

5-Acetyl-2,3-dihydro-
1,4-thiazine (roasty, popcorn)

2-Acetyltetrahydropyridine
(cracker, popcorn)

(Z)-2-propenyl-3,5-dimethyl-
pyrazine (roasted potato)

3-Thiazolidineethanethiol
(roasty, popcorn)

(E,E,Z)-2,4,6-nonatrienal
(flaked oats)

2-Furfuryl mercaptan
(coffee)

Furfuryl methyl disulfide
(mocha coffee)

3-Mercapto-3-methylbutyl
formate (roasted coffee)

Figure 9.5 Representative Maillard-generated and cereal-like character-impact flavor compounds.

Nut flavors

Pyrazines are the major compound classes in peanuts, formed by the thermally induced Maillard reaction (with the exception of methoxy pyrazines) [102]. Two pyrazines that represent peanut flavor character are 2,5-dimethylpyrazine (nutty) and 2-methoxy-5-methylpyrazine (roasted nutty); they are listed in Table 9.5.

Benzaldehyde has long been known as the character-impact compound of oil of bitter almond. It possesses an intense almondlike flavor in the context of savory applications; in sweet systems, it becomes cherrylike. 5-Methyl-2-thiophene-carboxaldehyde also provides almond flavor character, and occurs naturally in roasted peanuts [100].

The character-impact compound of hazelnuts, (E)-5-methyl-2-hepten-4-one (filbertone), undergoes isomerization during the roasting process [103]. Of the four possible geometric and enantiomeric isomers formed, all exhibit the typical hazelnut aroma, but the *trans*-(S)-isomer has the strongest impact. 5-Methyl-4-heptanone also contributes a fruity, hazelnut-like aroma. Structures are presented in Figure 9.6.

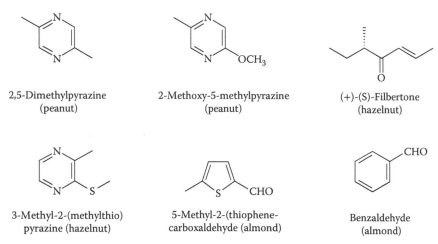

| 2,5-Dimethylpyrazine (peanut) | 2-Methoxy-5-methylpyrazine (peanut) | (+)-(S)-Filbertone (hazelnut) |
| 3-Methyl-2-(methylthio) pyrazine (hazelnut) | 5-Methyl-2-(thiophene-carboxaldehyde (almond) | Benzaldehyde (almond) |

Figure 9.6 Representative character-impact flavor compounds in peanuts and tree nuts.

Table 9.5 Character-Impact Flavor Compounds in Nuts

Character-impact compounds	CAS Registry number	Occurrence	Reference
Benzaldehyde	[100-51-6]	Almond	4
5-Methyl-2-thiophenecarboxal	[13679-70-4]	Almond	4, 100
γ-Nonalactone	[104-61-0]	Coconut	102
δ-Decalactone	[705-86-2]	Coconut	102
(E)-5-Methyl-2-hepten-4-one	[81925-81-7]	Hazelnut	103
5-Methyl-4-heptanone	[15726-15-5]	Hazelnut	103
2,5-Dimethylpyrazine	[123-32-0]	Peanut	4
2-Methoxy-5-methylpyrazine	[68358-13-5]	Peanut	4

As discussed in the section titled *Fruit flavors*, δ-lactones (e.g., δ-decalactone and δ-octalactone) possess a coconut flavor character. However, γ-nonalactone has the most intense coconut-like aroma as an individual character-impact compound, but it occurs only in artificial coconut flavors [102]. As the side-chain length increases, the character of γ-lactones changes to a peachlike aroma [14].

Meat and seafood flavors

Sulfur-containing heterocyclic compounds are associated with meaty characteristics. Two compounds with the most potent meaty characteristics include 2-methyl-3-furanthiol (threshold of 1 ppt) and the corresponding dimer bis-(2-methyl-3-furyl) disulfide (threshold of 0.02 ppt) [22]. Both substances have been identified in cooked beef and chicken broth, and they have a strong meaty quality upon dilution. 2-Methyl-3-furanthiol also occurs in canned tuna fish aroma [104]. The disulfide has a recognizable aroma character of "rich, aged beef, prime rib" [18]. Both compounds are produced from the thermal degradation of thiamin [105]. A related compound, 2-methyl-3-(methylthio)furan, is the character-impact compound for roast beef [22]. Other potent modifiers, such as 2-acetyl-2-thiazoline, impart a potent "roasty, popcorn" note, which enhances the meaty and roast flavor [106]. 2-Ethyl-3,5-dimethylpyrazine and 2,3-diethyl-5-methylpyrazine also contribute potent roasty notes to roast beef flavor [107]. 2-Acetyl-2-thiazoline was subsequently detected as a potent odorant of chicken broth and cooked chicken [108]. A summary of character-impact compounds for meat and seafood flavors is shown in Table 9.6.

A brothy compound associated with boiled beef 4-methyl-5-(2-hydroxyethyl)thiazole (sulfurol) is a "reaction flavor" product from the hydrolysis of vegetable protein. It is suspected that a trace impurity (2-methyltetrahydrofuran-3-thiol) in sulfurol is the actual "beef broth" character-impact compound [18]. Another reaction product flavor chemical is 2,5-dimethyl-1,4-dithiane-2,5-diol (the dimer of mercaptopropanone); it has an intense chicken-broth odor. 2-Pyrazineethanethiol, a synthetic pyrazine, provides excellent pork character [18].

Lipid components associated with meat fat, especially unsaturated aldehydes, play a significant role in species-characterization flavors. For example, (E,Z)-2,4-decadienal exhibits the character impact of chicken fat and freshly boiled chicken [108]. (E,E)-2,6-Nonadienal has been suggested as the component responsible for the tallowy flavor in beef and mutton fat [105]. 12-Methyltridecanal was identified as a species-specific odorant of stewed beef and provides a tallowy, beeflike flavor character [109]. While aldehydes provide desirable flavor character to cooked meat, they can contribute rancid and "warmed-over" flavors at high concentrations, resulting from the autoxidation of lipids [110].

Table 9.6 Character-Impact Flavor Compounds in Meat and Fish

Character-impact compounds	CAS Registry number	Occurrence	Reference
Dimethyl sulfide	[75-18-3]	Clam, oyster	111
Methional	[3268-49-3]	Boiled clam	117
		Crustaceans	118, 119
Pyrrolidino-2,4-dimethyl-dithiazine	[116505-60-3]	Roasted shellfish	120
2-Acetylthiazole	[24295-03-2]	Boiled clam	117
2-Acetyl-2-thiazoline	[22926-41-8]	Roast beef	106
		Cooked chicken	108
		Crustaceans	118, 119
4-Methylnonanoic acid	[45019-28-1]	Lamb	105
4-Methyloctanoic acid	[54947-74-9]	Lamb	105
2-Pentylpyridine	[2294-76-0]	Lamb	105
4-Methyl-5-(2-hydroxyethyl)thiazole	[137-00-8]	Roasted meat	18
2-Methyl-3-(methylthio)furan	[63012-97-5]	Roast beef	22
2-Ethyl-3,5-dimethyl-pyrazine	[13925-07-0]	Roast beef	107
2,3-Diethyl-5-methyl-pyrazine	[18138-04-0]	Roast beef	107
2-Methyltetrahydrofuran-3-thiol	[57124-87-5]	Brothy, meaty	18
2-Methyl-3-furanthiol	[28588-74-1]	Meat, beef	22
		Canned tuna fish	
Bis-(2-methyl-3-furyl)disulfide	[28588-75-2]	Aged, prime-rib	22
12-Methyltridecanal	[75853-49-5]	Beef, stewed	109
(E,E)/(E,Z)-2,4-Decadienal	[25152-84-5]	Cooked chicken	18, 108
2,5-Dimethyl-1,4-dithiane-2,5-diol	[55704-78-4]	Chicken broth	18
Pyrazineethanethiol	[35250-53-4]	Pork	18
(Z)-1,5-Octadien-3-one	[65767-22-8]	Salmon, cod	114
(E,Z)-2,6-Nonadienal	[557-48-2]	Trout, boiled	115
(Z,Z,Z)-5,8,11-Tetradecatrien-2-one	[85421-52-9]	Shrimp, cooked	116
2,4,6-Tribromophenol	[118-79-6]	Shrimp, ocean fish	121

Two fatty acids, 4-methyloctanoic and 4-methylnonanoic acid, provide the characteristic flavor of mutton [105]. 2-Pentylpyridine has been identified as the most abundant alkylpyridine isolated from roasted lamb fat. This compound has a fatty, tallowy aroma at an odor threshold of 0.6 ppb and is suspected to negatively impact the sensory acceptance of lamb and mutton [105].

The "fishy" aroma of seafood is incorrectly attributed to trimethyl amine. Flavor formation in fresh- and saltwater fish results from complex enzymatic, oxidative, and microbial reactions of ω-3 polyunsaturated fatty acid precursors (e.g., eicosapentaenoic acid) [111–113]. Hence, fish flavor is primarily composed of noncharacterizing "planty" or "melonlike" aromas from lipid-derived unsaturated carbonyl compounds. Examples are (Z)-1,5-octadien-3-one (geranium-like aroma) in boiled cod [114] and (E,Z)-2,6-nonadienal (cucumber-like aroma) in boiled trout [115]. 5,8,11-Tetradecatrien-2-one exhibits a distinct seafood aroma character described as "cooked shrimplike" or "minnow bucket" [116].

Three notable sulfur volatiles in boiled clam were determined by AEDA: (1) 2-acetyl-2-thiazoline (roasted aroma), (2) 2-acetylthiazole (popcorn aroma), and (3) methional (boiled potato aroma) [117]. Methional and 2-acetyl-2-thiazoline also contribute to the meaty and nutty/popcorn aroma notes in cooked crustaceans such as crab, crayfish, lobster, and shrimp [118,119]. An extremely potent odorant in cooked shellfish, including shrimp and clam, was identified as pyrrolidino[1,2-e]-4H-2,4-dimethyl-1,3,5-dithiazine [120]. This dithiazine contributes a roasted character to boiled shellfish and has the lowest odor threshold recorded to date, 10^{-5} ppt in water. 2,4,6-Tribromophenol and other bromophenol isomers have been associated with the ocean-, brine-, and iodine-like flavor character in seafood such as Australian ocean fish and prawns. The source of bromophenols is thought to be polychaete worms, which form an important part of the diet for many fish and prawn species [121]. Dimethyl sulfide is reported to be the character aroma of stewed clams and oysters [111]. Representative structures for meat and seafood flavor-impact compounds are shown in Figure 9.7.

Cheese and dairy flavors

With a few exceptions, many of the known important flavors in dairy products do not provide characterizing roles. This is especially true for milk, cheddar cheese, and cultured products, such as sour cream and yogurt. δ-Lactones are important flavors in butter, buttermilk, and cheeses, which are derived from triglycerides containing hydroxyl fatty acids. Although they do not directly contribute as dairy character-impact compounds, lactones play key supporting roles in dairy flavors. The subject of key odor-active compounds in milk and dairy flavors has been recently reviewed [122–126]. A summary of character-impact compounds for cheese and dairy flavors is presented in Table 9.7.

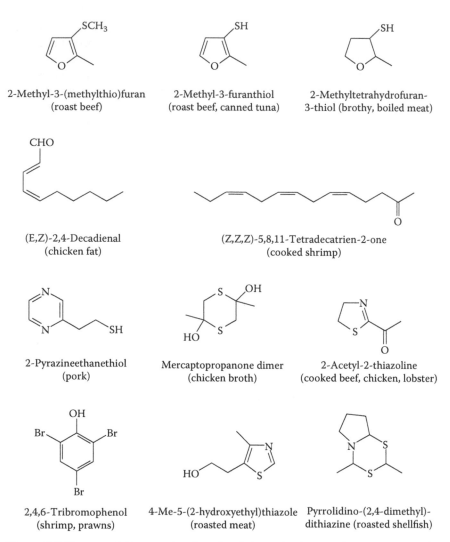

Figure 9.7 Representative meat and seafood character-impact flavor compounds.

Urbach [127] discusses the formation of volatile flavor compounds in different varieties of cheeses and provides a compilation of important aroma compounds. A recent qualitative assessment by Sable and Cottenceau [128] surveys the significant flavor volatiles that have been identified in soft mold-ripened cheeses including Camembert, Brie, blue, Gorgonzola, Munster, and Limburger among others. Octen-3-ol, 2-phenylethanol, and 2-phenylethyl acetate are character-impact components in Camembert-type cheese; these compounds together with sulfur compounds, 1-octen-3-one, and δ-decalactone are reported to be the key

Table 9.7 Character-Impact Flavor Compounds in Cheese and Dairy Products

Character-impact compounds	CAS Registry number	Occurrence	Reference
2,3-Butanedione	[431-03-8]	Butter	141, 146
δ-Decalactone	[705-86-2]	Butter	141, 145, 146
6-Dodecen-γ-lactone	[156318-46-6]	Butter	141, 145
		Cheese, cheddar	134, 135
2-Heptanone	[110-43-0]	Cheese, blue	127
1-Octen-3-ol	[3391-86-4]	Cheese, Camembert	128
Butryric acid	[107-92-6]	Cheese, cheddar	134
2-Acetyl-2-thiazoline	[22926-41-8]	Cheese, cheddar	134, 135, 138
Tetramethylpyrazine	[1124-11-4]	Cheese, cheddar	138
Methional	[3268-49-3]	Cheese, cheddar	134
		Cheese, Parmigiano Reggiano	131–133
Skatole	[83-34-1]	Cheese, cheddar	134, 138
2-Acetyl-1-pyrroline	[85213-22-5]	Cheese, cheddar	134, 138
		Whey protein concentrate	144
2,6-Dimethylpyrazine	[108-50-9]	Parmigiano Reggiano	132
		Whey, dried	143
4-Mercapto-4-methyl-2-pentanone	[19872-52-7]	Cheese, cheddar	139
4-Mercapto-3-methyl-2-pentanone	[385766-53-0]	Cheese, cheddar	139
Homofuraneol	[110516-60-4]	Cheese, Swiss	147
		Cheese, cheddar	134, 135
Propionic acid	[79-09-4]	Cheese, Swiss	147
Ethyl-3-mercapto-propionate	[5466-06-8]	Camembert, Munster	129
4-Methyloctanoic acid	[54947-74-9]	Cheese, goat	130
4-Ethyloctanoic acid	[16493-80-4]	Cheese, goat	130
(Z)-4-Heptenal	[6728-31-0]	Cream	4
		Cheese, cheddar	135
(E,E)-2,4-Nonadienal	[5910-87-2]	Cream	122
1-Nonen-3-one	[24415-26-7]	Yogurt, milk	148
Furaneol	[3658-77-3]	Milk, nonfat dry	142
		Whey protein concentrate	143, 144
		Butter, heated	145, 146

aroma substances for Camembert. A thiolester ethyl 3-mercaptopropionate was reported for the first time in Munster and Camembert cheeses [129]. This sulfur volatile was described at low concentrations as having pleasant "fruity, grapy, rhubarb" characters. 4-Methyl- and 4-ethyloctanoic acids contribute the "waxy/animal" character in fresh chèvre-style goat cheese flavor [130]. 2-Heptanone, 2-nonanone, and short- and moderate-chain fatty acids are the dominant character compounds of blue cheese flavor. Sulfur compounds, especially methanethiol, hydrogen sulfide, and dimethyl disulfide, contribute to the strong garlic/putrid aroma of soft-smear or surface-ripened cheeses. Key aroma compounds in Parmigiano–Reggiano cheese were recently reported, including 3-/2-methylbutanal, 2-methylpropanal, dimethyltrisulfide, diacetyl, methional, ethyl C_4–C_8 fatty acid esters, and C_2–C_8 fatty acids [131–133]. In the basic fraction, 2,6-dimethylpyrazine and 6-ethyl-2,3,5-trimethylpyrazine had very strong nutty, baked characters and the highest flavor intensities among the six pyrazines detected [132].

By a wide margin, cheddar is the most popular cheese flavor in North America. Although its flavor is described as "sweet, buttery, aromatic, and walnut," there is no general consensus among flavor chemists about the identity of individual compounds or groups of compounds responsible for cheddar flavor. At present, it is thought to arise from a unique balance of key volatile components rather than a unique character-impact compound. Sensory-guided flavor studies concluded that butyric acid, acetic acid, methional, homofuraneol (5-ethyl-4-hydroxy-2-methyl-(2H)-furan-3-one), (E)-2-nonenal, (Z)-4-heptenal, (Z)-1,5-octadien-3-one, and 2-acetyl-1-pyrroline are primary contributors to the pleasant mild flavor of cheddar cheese [134,135]. Important contributors to cheddar aroma are 2,3-butanedione, dimethyl sulfide, dimethyl trisulfide, and methanethiol [134]. Two recent studies confirmed that concentrations of hydrogen sulfide and dimethyl disulfide increased only during the initial stages of cheddar cheese aging, whereas those of dimethyl sulfide, dimethyl trisulfide, and methanethiol continued to increase throughout the aging process [136,137]. A desirable nutty flavor supports "sulfur" and "brothy" characters in a quality aged cheddar cheese sensory profile. 2-Acetyl-2-thiazoline [134] and 2-acetyl-1-pyrroline [134,135] contribute "roasted, corny" flavors, which were suggested to be related to the nutty cheddar flavor. In an effort to characterize the nutty flavor in cheddar cheese, a comprehensive sensory–analytical study concluded that Strecker aldehydes (2-methylpropanal; 2- and 3-methylbutanal), which individually are "green, malty, chocolate," provide an enhanced overall nutty flavor when added to cheddar cheese [138]. Novel polyfunctional thiols were recently reported in aged cheddar cheese. Among these were 4-mercapto-4-methyl-2-pentanone (catty note) and 4-mercapto-3-methyl-2-pentanone (cooked milk, sweet note) [139]. In the case of cheddar cheese powders, dimethyl sulfide imparts a desirable "creamed corn" flavor [140].

The most aroma-active compound in fresh, sour-cream butter was reported as the character-impact compound diacetyl, with supporting contributions from δ-decalactone, (Z)-6-dodeceno-γ-lactone, and butyric acid [141].

Key aroma-active compounds have been reported in dried dairy products including nonfat milk and whey powders. Furaneol is the primary contributor to the sweet, caramelized aroma character of nonfat dry milk, with supporting contributions from heat-generated sotolon, maltol, vanillin, and (E)-4,5-epoxy-(E)-2-decenal [142]. Compounds important to dried sweet whey from cheddar cheese manufacture include butyric and hexanoic acids; maltol; Furaneol; dimethylpyrazines; 2,3,5-trimethylpyrazine; and several lactones [143]. Additional flavor-impact compounds identified in whey protein concentrate and whey protein isolate were butyric acid (cheesy note); 2-acetyl-1-pyrroline (popcorn note); 2-methyl-3-furanthiol (brothy/burnt note); Furaneol (maple/spicy note); 2-nonenal (fatty note); (E,Z)-2,6-nonadienal (cucumber note); and (E,Z)-2,4-decadienal (fatty/oxidized note) [144].

A summary of potent flavor compounds in dairy products is presented in Table 9.7. Representative structures of significant dairy flavorants including homofuraneol in heated butter [145,146], mild cheddar cheese [134], and Swiss cheese [147]; δ-decalactone in butter and buttermilk [141]; (Z)-6-dodecen-γ-lactone in butter [145,146] and mild cheddar cheese [134,135]; and 1-nonen-3-one in milk and yogurt [148] are shown in Figure 9.8.

Characterizing aromas in off-flavors

Flavor defects, so-called taints, malodors, or off-flavors, are sensory attributes that are not associated with the typical aroma and taste of foods and beverages. These defects can range from subtle to highly apparent and are often significant detractors to food quality. Off-flavors can be produced due to several possible factors: contamination (air, water, packaging, or shipping materials); ingredient mistakes in processing; or generation (chemical or microbial) in the food itself. In the latter instance, generation of off-flavors in foods may result from oxidation, nonenzymatic browning, chemical reactions between food constituents, light-induced reactions, microbial fermentations, or enzymatic pathways.

Over the past 15 years, numerous complete and detailed reviews have discussed the occurrence of off-flavors in food and packaging systems [149–153]. The intent of this section is to summarize recent highlights and off-flavorants of significance, without striving for comprehensiveness.

Lipid-derived volatile compounds play an important role in many food flavors. These compounds contribute to the characteristic and desired flavor attributes of foods, but they can also cause off-flavors depending on their concentrations relative to other sensorially relevant odorants.

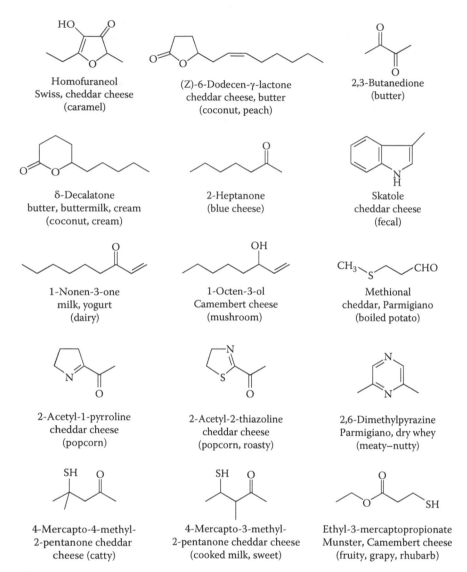

Homofuraneol
Swiss, cheddar cheese
(caramel)

(Z)-6-Dodecen-γ-lactone
cheddar cheese, butter
(coconut, peach)

2,3-Butanedione
(butter)

δ-Decalatone
butter, buttermilk, cream
(coconut, cream)

2-Heptanone
(blue cheese)

Skatole
cheddar cheese
(fecal)

1-Nonen-3-one
milk, yogurt
(dairy)

1-Octen-3-ol
Camembert cheese
(mushroom)

Methional
cheddar, Parmigiano
(boiled potato)

2-Acetyl-1-pyrroline
cheddar cheese
(popcorn)

2-Acetyl-2-thiazoline
cheddar cheese
(popcorn, roasty)

2,6-Dimethylpyrazine
Parmigiano, dry whey
(meaty–nutty)

**4-Mercapto-4-methyl-
2-pentanone cheddar
cheese (catty)**

**4-Mercapto-3-methyl-
2-pentanone cheddar cheese
(cooked milk, sweet)**

Ethyl-3-mercaptopropionate
Munster, Camembert cheese
(fruity, grapy, rhubarb)

Figure 9.8 Representative cheese and dairy character-impact flavor compounds.

For example, the "cardboard" off-flavor of butter oil is primarily related to the presence of (E)-2-nonenal, which is formed by the autoxidation of palmitoleic acid [154]. Carbonyl compounds formed by lipid peroxidation were identified in cooked beef, which develops a warmed-over flavor from reheating after 2 days of refrigerated storage [110]. Warmed-over flavor is principally caused by the formation of (E)-4,5-epoxy-(E)-2-decenal (metallic note) and hexanal (green note), which are not present in freshly cooked beef. Similarly, for boiled chicken "green/cardboardlike/metallic" off-odors were formed during refrigerated storage and reheating, primarily from a sevenfold increase in hexanal [108]. These and other unsaturated carbonyl compounds, including (E,E)-2,4-decadienal (deep-fried note) and 1-octen-3-one (mushroomlike/beany note), are primarily responsible for the rancid off-flavors in soybean and canola oils, which are caused by the oxidation of linoleic and linolenic acids [155]. The light-induced compound 3-methyl-2,4-nonanedione (strawy, lard-like, beany note) strongly contributes to the reverted off-flavor of oxidized soybean oil, because of its extremely low odor threshold (0.007–0.014 ppt) [156]. Melon odors are associated with foods cooked in partially hydrogenated soybean oils that have undergone oxidative deterioration during heating. 6-Nonenal (cucumber/melon note) from (Z,Z)-9,15-linoleic acid is reported to be a character flavor associated with these deteriorated oils [157]. (Z)-3-Hexenal and (Z,Z)-3,6-nonadienal were shown to contribute substantially to the fatty, "fishy" off-flavor of boiled trout, which was in frozen storage for several months before cooking [115]. Other carbonyl compounds that are likely to contribute to characteristic "fishy" off-flavors in oxidized seafood because of their low odor thresholds include 2,6-nonadienal and 1,5-octadien-3-one [111]. A summary of off-flavor impact compounds for foods, beverages, and packaging materials is presented in Table 9.8, and the representative chemical structures are given in Figure 9.9.

The characteristic odor-active components in cardboard packaging material are *cis-* and *trans-*2-nonenal, along with an unidentified "woody smelling" constituent [158]. *trans*-2-Nonenal is considered the characteristic volatile responsible for a "stale, cardboard" off-flavor in aged packaged beer. Mechanistic studies using labeled nonenal confirmed that cardboard off-flavor in finished beer arises from lipid autoxidation during wort boiling and not from lipoxygenase activity during the mashing step [159]. Studies also revealed that 2-furfuryl ethyl ether is responsible for an astringent, stale off-flavor in beer [160]. The stale flavor was reproduced by adding furfuryl ether and *trans*-2-nonenal to fresh beer, but not by adding either compound individually. Recently, 4-vinylsyringol (3,5-dimethoxy-4-hydroxystyrene) was identified as a contributor to stale beer odor in naturally aged beer [161]. Its odor character from GC–O was described as "smoky, tobacco, old beer." The chemically similar 4-vinylguaicol contributes "phenolic, medicinal, or clove" off-flavors at higher levels from wheat beers brewed with certain yeast strains [162]. Lipid-derived (E,E)-2,4-decadienal, in addition to hexanal, (E)-2-octenal, and (E)-2-nonenal, was shown to be the most potent off-flavor compound in precooked vacuum-packaged potatoes [163]. In dry raw spinach, (Z)-1,5-octadien-3-one and methional are responsible for a "fishy" off-flavor when they are present in a 1:100 ratio. 3-Methyl-2,4-nonanedione produces a "haylike" off-flavor character from the oxidative degradation of furanoid fatty acids in dry spinach [164] and in dry parsley [165].

Table 9.8 Off-Flavor Impact Compounds in Food and Beverage Products

Impact compounds	CAS Registry number	Off-flavor descriptors	Occurrence	Reference
Geosmin	[19700-21-1]	Musty, earthy	Catfish, wheat, water	185, 192
2,4,6-Trichloroanisole	[87-40-1]	Musty, moldy	Coffee, wine corks, raisins	191, 192, 188
2,4,6-Tribromoanisole	[607-99-8]	Musty, moldy	Casein, wine corks, wine	189, 187
2-Methylisoborneol	[2371-42-8]	Earthy, musty	Coffee, catfish, beans	191, 185, 192
2-iso-Propyl-3-methoxypyrazine	[25773-40-4]	Peasy	Coffee, cocoa beans	191, 151
Ethyl formate	[109-94-4]	Fermented	Coffee, roasted	184
4,4,6-Trimethyl-1, 3-dioxane	[1123-07-5]	Musty	Packaging film	195
8-Nonenal	[39770-04-2]	Smoky, plastic	Polyethylene packaging	196
Iodocresol	[16188-57-1]	Medicinal	Lemon cake mix	194
(E)-2-Nonenal	[2463-53-8]	Cardboard, stale	Beer, packaged	166
		Cardboard	Oxidized butter	154
		Cardboard	Soybean lecithin	155
		Cardboard	Cardboard packaging	158
(E)-4,5-Epoxy-(E)-2-decenal	[134454-31-2]	Metallic	Warmed over (beef)	110
		Metallic	Oxidized soybean oil	155
		Metallic	Butter	78
(E,Z)-2,6-Nonadienol	[7786-44-9]	Metallic	Buttermilk	205
1-Octen-3-one	[4312-99-6]	Metallic, mushroom	Butterfat	200
		Beany	Rancid soybean oil	155
Hexanal	[66-25-1]	Green grass	Rancid soybean oil	155
		Rancid	Warmed over (beef)	110
(E,E)-2,4-Decadienal	[25152-84-5]	Deep fried	Rancid soybean oil	155

Compound	CAS	Off-flavor	Source	Reference
2,3-Diethyl-5-methylpyrazine	[18138-04-0]	Roasty, earthy	Soybean lecithin	155
6-Nonenal	[6728-35-4]	Cucumber, melon	Partially hydrogenated soy oil	157
(Z)-3-Hexenal	[6789-80-6]	Fatty, fishy	Aged trout	115
(Z,Z)-3,6-Nonadienal	[21944-83-2]	Fatty, fishy	Aged trout	115
5α-Androst-16-en-3-one	[18339-16-7]	Urine	Boar meat	151, 213
Skatole	[83-34-1]	Fecal like	Boar meat, potato chips	213, 151
		Medicinal	Beef	214
2-Furfuryl ethyl ether	[6270-56-0]	Stale, astringent	Beer	160
3-Methyl-2-butene-1-thiol	[5287-45-6]	Skunky	Beer (light-struck)	166, 167
S-Methyl hexanethioate	[2432-77-1]	Cabbagy, rubbery	Beer	169
4-Vinylsyringol	[28343-22-8]	Old beer, tobacco, smoky	Beer	161
4-Vinylguaiacol	[7786-61-0]	Medicinal, clove	Beer, wort	162
		Rotten, old fruit	Orange juice, apple juice	216, 217
4-Mercapto-4-methyl-2-pentanone	[19872-52-7]	Cat urine, ribes	Beer	170
Methional	[3268-49-3]	Worty	Beer (alcohol-free)	176
		Cooked vegetables	Oxidized white wine, orange juice	172, 173, 174
		Potato	UHT-processed milk	178, 210, 211
Dimethyl disulfide	[624-92-0]	Sunlight off-flavor	Milk, skim	179
(E)-1,3-Pentadiene	[504-60-9]	Kerosene	Cheese (sorbic acid)	151, 206

(Continued)

Table 9.8 Off-Flavor Impact Compounds in Food and Beverage Products (*Continued*)

Impact compounds	CAS Registry number	Off-flavor descriptors	Occurrence	Reference
2,4,5-Trimethyloxazole	[20662-84-4]	Melon, kiwifruit	Dried sour cream	140
Tetradecanal	[124-25-4]	Sickening, aldehydic	Milk powder	207
β-Ionone	[79-77-6]	Haylike	Milk powder	207
Benzothiazole	[95-16-9]	Sulfuric, quinoline	Milk powder	207
2,6-Dimethylpyrazine	[108-50-9]	Cooked milk	UHT-processed milk	210
2-Ethyl-3-methylpyrazine	[15707-23-0]	Cooked milk	UHT-processed milk	210
2-Aminoacetophenone	[551-93-9]	Gluey, glutinous	Milk powder, casein	197, 198, 208
		Untypical aging (UTA)	White wine	199
3-Methyl-2, 4-nonanedione	[113486-29-6]	Strawy, beany	Soybean oil (light-induced)	156
		Haylike	Dried spinach, parsley	164, 165
(Z)-1,5-Octadien-3-one	[65767-22-8]	Fishy	Dried spinach, old fish	164, 111
Bis(2-methyl-3-furyl)disulfide	[28588-75-2]	Vitamin B_1 odor	Thiamin degradation	203
4-Methyl-2-isopropylthiazole	[15679-13-7]	Vitamin, cabbage	Orange juice (Vitamin B_2)	180
Sotolon	[28664-35-9]	Burnt, spicy	Citrus soft drink	202
2-Methyl-3-furanthiol	[28588-74-1]	Aged flavor	Orange juice	174, 175
(S)-(+)-Carvone	[2244-16-8]	Woody, terpeny	Oxidized orange juice	215

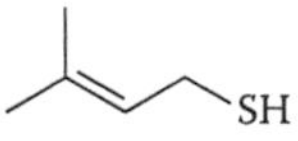

3-Methyl-2-butene-1-thiol
light-struck beer

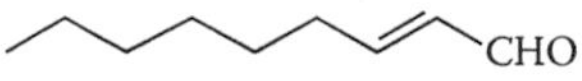

(E)-2-Nonenal
cardboard (beer, butter,
packaging)

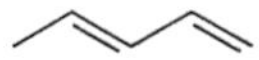

1,3-Pentadiene
kerosine (sorbate/cheese)

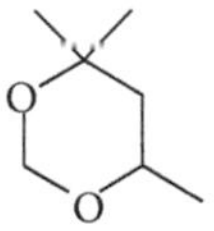

4,4,6-Trimethyl-1,3-dioxane
musty packaging film

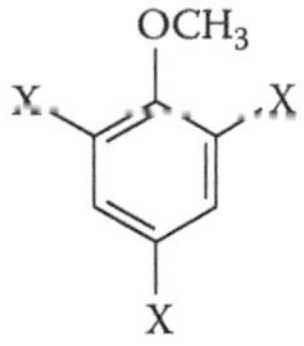

X=Cl 2,4,6-Trichloroanisole
X=Br 2,4,6-Tribromoanisole
musty (coffee, wine corks, casein)

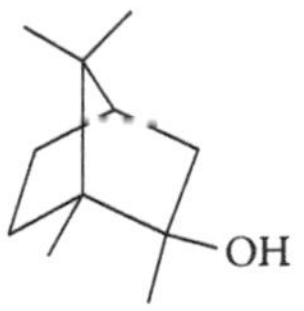

2-Methylisoborneol
earth–musty (coffee, catfish)

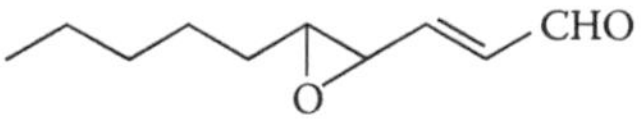

trans-4,5-Epoxy-(E)-2-decenal
metallic (warmed-over
flavor beef; soybean oil)

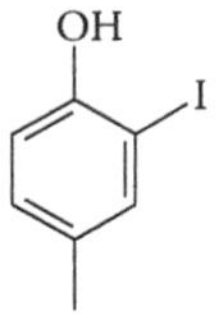

Iodocresol
medicinal (lemon cake mix)

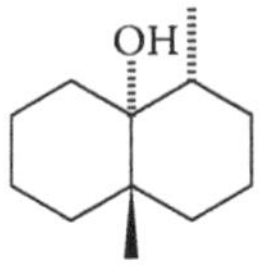

Geosmin
musty–earthy (fish, water)

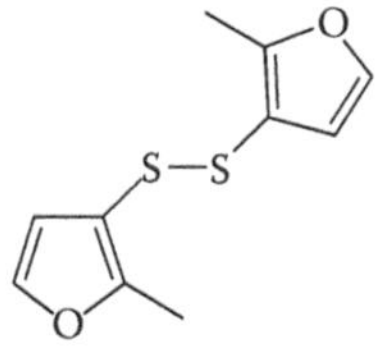

Bis-(2-methyl-3-furyl)disulfide
vitamin B$_1$ (thiamin)

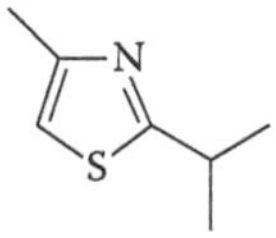

4-Methyl-2-iso-propylthiazole
vitamin B$_2$ (fortified orange juice)

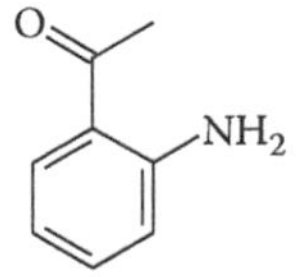

2-Aminoacetophenone
gluey (milk, wine)

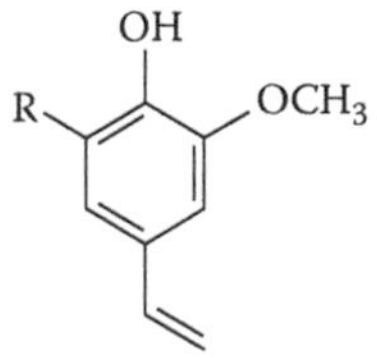

R=H 4-Vinylguaiacol
rotten (orange); clove (beer)
R=OCH$_3$ 4-Vinylsyringol
old beer, tobacco, smoky (beer)

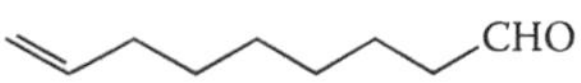

8-Nonenal
smoky plastic (oxidized
polyethylene)

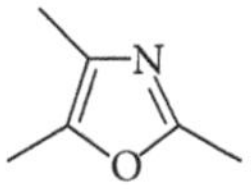

2,4,5-Trimethyloxazole
melonlike (dried cream)

Figure 9.9 Representative off-flavor character-impact compounds.

Exposure of beer to light has been shown to produce 3-methyl-2-butene-1-thiol, which produces a skunky off-flavor in "sun-struck" or "light-struck" ales [166,167]. This mercaptan has a sensory threshold of 0.05 ppb in beer. It results from complex light-induced degradations of isohumulones (hop-derived bitter *iso*-acids) to form free-radical intermediates, which subsequently react with the thiol group of cysteine. The light-struck off-flavor can be controlled in beer through the use of improved packaging technology (colored glass bottles), chemically modified hop bitter acids, or antioxidants or its precipitation with high-molecular-weight gallotannins by the addition of zinc salts [168]. In addition to dimethyl sulfide, thioesters have been reported to contribute a cabbagy, rubbery off-note; they sometimes are derived from hops in beer, the most significant being *S*-methyl hexanethioate, which has a detection threshold of 1 ppb [169]. Diacetyl can produce an undesirable buttery off-character in beer through accelerated fermentation, whereby brewer's yeast does not convert all the diketone intermediates to flavor-inactive acetoin and 2,3-butanediol [166]. Another staling off-flavor in beer is described as catty/ribes, whose character comes from 4-mercapto-4-methyl-2-pentanone (cat ketone). This off-flavor can develop rapidly in beers that are packaged and stored with a high air content in the headspace [170]. A similar catty/ribes off-flavor in aged lager beer was attributed to 3-mercapto-3-methylbutyl formate [171].

Strecker aldehydes are a frequent source of off-flavors in fermented products. Development of off-flavors in oxidized white wines typically marks the end of shelf life for these wines. Methional (3-methylthiopropionaldehyde) was identified as producing a "cooked vegetables" off-flavor character in a young white wine that had undergone spontaneous oxidation [172]. Methional levels increased in wines spiked with methionol or methionine, suggesting its formation via direct peroxidation or Strecker degradation of methionine. The importance of methional in the development of characteristic oxidation notes in white wine was further demonstrated [173]. Methional and 2-methyl-3-furanthiol are purported off-flavor components in stored orange juice [174]. Methanethiol, 1-*p*-menth-1-ene-8-thiol, 2-methyl-3-furanthiol, and dimethyl trisulfide contributed atypical "tropical fruit/grapefruit" character to canned orange juice [175]. Methional was shown to impart a worty off-flavor in alcohol-free beer, with more sensory significance than was previously attributed to 3-methyl- and 2-methylbutanal for this off-taste [176]. In an alcohol-free beer medium, perception of worty off-flavors is strengthened by the absence of ethanol and by the higher level of sugars. Methional, methanethiol, and dimethyl sulfide are the sources of sulfur and cooked off-flavors in ultrahigh-temperature (UHT)–processed soy milks [177]. Similar to the off-flavor in UHT-processed cow's milk, 2-acetyl-1-pyrroline, 2-acetylthiazole, and carbonyl compounds also contributed to the cooked flavor impact [178]. Methional was initially thought to be responsible for the "sunlight" off-flavor in photooxidized milk exposed to high-intensity fluorescent light or sunlight. This card-boardlike off-flavor is generated via photooxidation of methionine in the presence of riboflavin as a sensitizer. More recently, dimethyl disulfide was identified as the source of the light-induced off-flavor in skim milk [179]. Another vitamin-derived off-odor problem was recently described in which a pineapple fruit juice beverage was fortified with riboflavin. The "vitamin, cabbage, brothy, vegetable soup" off-odor was characterized as 4-methyl-2-isopropylthiazole, which resulted from riboflavin-sensitized Strecker degradation of valine, cysteine, and methionine followed by reaction of the resulting aldehydes with ammonia and hydrogen sulfide [180].

Sulfur compounds present in wine can have a detrimental effect on its aroma character, producing odors described as "garlic, onion, and cauliflower" or the so-called Boeckser aromas. This sulfurous character is correlated with 2-methyl-3-hydroxythiophene, 2-methyl-3-furanthiol, and ethanethiol, and their concentrations in wine are influenced by winery practices and the use of certain winemaking yeasts [181]. Off-flavors in European wines were associated with the nonvolatile bis(2-hydroxyethyl) disulfide, a precursor to the "poultrylike" character of 2-mercaptoethanol and hydrogen sulfide [182].

Other undesirable "Brett" off-flavors are produced in wines due to contamination in grape musts and wineries with *Brettanomyces* yeast strains, named after their original discovery in British beers. These off-flavors are typically described as "Band-Aid," "barnyard," "horse sweat," "plastic," and "wet animal." A recent study identified the key character compounds in high-Brett wines as isovaleric acid, 4-ethyl guaiacol, and 4-ethyl phenol; however, a potent, plastic note was not characterized in this study [183].

The quality of coffee is influenced by many factors including its variety, agricultural and postharvest conditions, roasting parameters, and brewing. Recently, ethyl formate was identified as being responsible for the characteristic "fermented" off-note, which is occasionally perceived in the aroma of roasted and ground coffee [184]. As this fermented off-flavor is a well-known issue for coffee growers and the coffee trade, ethyl formate is useful as a quality marker for coffee.

Musty aromas and flavors are a major problem in a variety of foods and packaging materials, and many of the causative character-impact materials show powerful impact [185]. For example, 2,4,6-trichoroanisole is a highly odorous metabolite of a fungus that attacks wood, paperboard packaging, and wine corks [150,151]. Trichloroanisole has a musty, haylike odor and possesses an extremely low odor threshold (~0.05 ppt). Tricholoroanisoles, tribromoanisoles, and their corresponding phenols can provide musty off-flavor character to foods [186], wine corks/wine [187], process water, and raisins [188]. 2,4,6-Tribromoanisole was recently reported to impart a musty flavor defect to calcium caseinate and is formed from the fungal conversion of 2,4,6-tribromophenol contaminants in wood freight containers and plastic slip sheets [189]. It was present in tainted caseinate at concentrations of 2–560 ppb, with a sensory threshold of 20 ppt in 10% aqueous solution. About 20% of Brazilian coffee production exhibits the so-called Rio defect, which is characterized by a strong off-flavor that is often described as "medicinal, phenolic, or iodine-like" [190]. Occasionally, this defect also occurs in coffees of other origins. 2,4,6-Trichloroanisole in concentrations ranging from 1 to 100 ppb was identified as the most likely key compound for the Rio off-flavor as analyzed by capillary GC, GC–sniffing, and GC–MS. 2,4,6-Trichlorophenol, the probable precursor, was also found in most of these samples. Adding trichloroanisole to freshly brewed coffee imparted to it the same off-flavor notes as described in actual Rio coffee. Its perception threshold in coffee brew was found to be 8 ppt by aroma and 1–2 ppt for taste [190]. In Robusta coffees, 2-methylisoborneol provides an "earthy, tarry" character at 5 ppt and it must be removed during processing to approach the flavor of arabica coffees [191]. Geosmin (1,10-dimethyl-9-decalol) imparts an earthy musty off-odor to drinking water and a "muddy" odor to catfish and tilapia fish, which live in brackish water [185,192,193]. It is produced from *Actinomycetes* microorganisms in soil, planktonic algae, and fungi. 2-Methylisoborneol also provides an earthy musty character to catfish [192]. A medicinal off-flavor was produced in a lemon-flavored cake mix by the reaction between two minor ingredients of the cake, *p*-cresol from the lemon flavor and iodine from iodized salt. The resulting iodocresol was shown to possess a medicinal character at 0.2 ppb aroma threshold [194]. McGorrin, Pofahl, and Croasmun [195] studied a musty off-odor in printed plastic film, which was not a phenol as expected but was identified as 4,4,6-trimethyl-1,3-dioxane. It was formed during film manufacture as a reaction product of 2-methyl-2,4-pentanediol, a solvent coating to facilitate ink adhesion, and formaldehyde, a component of the ink.

Another common source of packaging-derived off-flavors is polyethylene packaging materials, which occasionally transfer undesirable off-flavors to foods stored in them. So-called plastic or "smoky-poly" off-flavors can be generated by minor plasticizer components in the resin or by oxidation of the polyolefin during polymerization and subsequent thermal extrusion into films, sheets, or containers. Polyethylene films that are manufactured under high-temperature conditions are more likely to contribute such off-flavors. Recently, a previously unidentified, potent plastic off-flavorant in oxidized polyethylene films was identified as 8-nonenal, which contributes a characteristic plastic off-flavor to corn chips packaged in the films [196].

Compounds that have positive character impact in some foods can become off-flavors in different food contexts. For example, indole and skatole in cheddar cheese flavor become fecal in the context of potato chips [151]. 2-Aminoacetophenone, having foxy character in Concord grape, imparts a "gluey" flavor in milk powder and casein through the degradation of tryptophan [197,198]. In white wines (such as Riesling), it contributes an "untypical aging" (UTA) note, presumably from oxidative degradation of the tryptophan metabolite indole-3-acetic acid after the sulfuration step in wine making [199]. 1-Octen-3-one (mushroom) imparts an undesired metallic flavor in dairy products and oxidized vegetable oils [200]. A recent study identified two key metallic odorants generated in aqueous ferrous sulfate solutions in which 1-nonen-3-one provides a 10-fold more potent metallic-smelling character relative to 1-octene-3-one [201]. Sotolon, the furanone character-impact flavor of fenugreek (Table 9.1) was recently shown to cause a "burnt, spicy" off-flavor in citrus soft drinks, generated by the reaction of ascorbic acid with ethanol [202]. Whereas 2-isopropyl-3-methoxy pyrazine contributes to pea an earthy potato flavor character, it provides an undesirable "peasy" off-flavor to Rwandan coffee [191] and fermented cocoa beans [151]. Although bis(2-methyl-3-furyl) disulfide contributes a desirable aged, prime-rib flavor in beef, it is the principal "vitamin B" off-odor resulting from thiamin degradation [203].

Character compounds that contribute to positive flavor impact at low levels can become off-flavors when they occur at higher concentrations. For example, dimethyl sulfide provides an appropriate "cornlike" background character to beer flavor at low levels, whereas it contributes a highly undesirable cooked vegetable or cabbagelike malodor to beer when present at levels significantly above its sensory threshold (30–45 ppb) [204]. Similarly, for cheddar cheese flavor it imparts a rotten vegetable taste when present at high levels [135].

Off-flavors in dairy products have been reviewed much in the literature [151,152]; however, there have been several recent developments. In sour-cream buttermilk, the key odorant responsible for a metallic off-flavor was identified as (E,Z)-2,6-nonadienol [205]. During cream fermentation, its formation occurs from the peroxidation of α-linolenic acid to generate the 2,6-nonadienal precursor, with subsequent reduction to dienol by starter culture reductases that remain active during storage. Metallic off-flavors are not formed readily in fermented sweet-cream buttermilk due to the presence of significantly lower concentrations of α-linolenic acid [205]. The common use of sorbic acid or potassium sorbate as a mold inhibitor in commercial dairy products often produces an off-flavor described as "kerosene, plasticlike, or paintlike," which may incorrectly be attributed to packaging materials. The source of the taint is (E)-1,3-pentadiene, which results from the decarboxylation of sorbates by lactic acid bacteria in yogurt, cheese, and margarine [151,206]. A "melon, ripe kiwifruit" off-flavor in spray-dried cultured dairy products was identified as 2,4,5-trimethyloxazole, a product of the Maillard reaction between diacetyl and arginine (whey source) or acetaldehyde and ammonia [140].

Other significant dairy taints studies include studies on the characteristic off-flavor in spray-dried skim milk powder, which was related to contributions from tetradecanal (sickening, aldehydic note), β-ionone (haylike note), and benzothiazole (sulfuric, quinoline note) at low-ppb levels [207]. The cause of an intense "musty, stale" off-flavor was identified as 2-aminoacetophenone in micromilled milk powder [208]. A recent AEDA study of the characteristic aroma components of dried rennet casein indicated that although 2-aminoacetophenone is a potent odorant, sensory reconstitution confirmed that hexanoic acid, indole, guaiacol, and *p*-cresol were the principal factors behind rennet casein's typical "animal/wet dog" odor [209]. In three studies of UHT-processed milk, the "UHT-milk-flavor" character is contributed by methional, δ-decalactone, 2-acetyl-1-pyrroline, and 2-acetylthiazole; 2,6-dimethylpyrazine; 2-ethylpyrazine; and 2-ethyl-3-methylpyrazine were also significant contributors [178,210]. Additionally, dimethyl sulfide, diacetyl, 2-heptanone, 2-nonanone, 2-methylpropanal, 3-methylbutanal, nonanal, and decanal were reported as significant flavor components of UHT milk off-flavor [211]. Formation of the UHT off-flavor was shown to be inhibited by the addition of 0.1% epicatechin during milk processing [178]. Fruity (pineapplelike) off-flavors in pasteurized milk indicate the presence of high levels of ethyl butyrate, ethyl hexanoate, ethyl octanoate, and ethyl decanoate esters, whereas rancid, soapy tastes arise from the presence of decanoic and dodecanoic acids [212].

As previously discussed in the section titled *Meat and seafood flavors*, lipid oxidation is generally related to flavor deterioration in meat and meat products. However, "boar taint," an intense urinelike off-odor, is attributed primarily to the flavor synergy between two compounds in boar fat, androstenone (5-α-androst-16-ene-3-one) from testes and the "fecal-like" skatole (3-methyl indole) from tryptophan breakdown [213]. Generally, women tend to be more sensitive to the odor than men. Skatole has also been implicated in a medicinal off-odor in beef [214].

Off-flavors in citrus oils such as orange oil have been related to autoxidation of limonene to (S)-carvone (caraway; Figure 9.1) and carveol, producing "woody, turpeny" off-flavors [215]. Other off-flavors in orange juice arise from the presence of 4-vinylguaiacol, which contributes an "old fruit" and "rotten" character via decarboxylation of ferulic acid [216]. Conversion of ferulic acid to 4-vinylguaiacol by yeast contaminants, with corresponding off-flavor development, has been reported in unpasteurized apple juice [217] and in beers and worts [162]. Thermal abuse during processing or high-temperature storage of orange and grapefruit juices produces Furaneol, which contributes a "sweet, pineapple" defect [216].

Conclusion

The objective of this chapter is to provide an updated review of character-impact compounds in flavors and off-flavors. Particular emphasis is placed on compounds that have been identified in natural flavor systems.

The summarized data can be applied in creative flavor-compounding efforts to replicate and monitor the production of flavors and in food processing to ensure flavor quality.

References

1. Nijssen, L. M., C. A. Visscher, H. Maarse, L. C. Willemsens, and M. H. Boelens, eds. 1996. *Volatile Compounds in Food. Qualitative and Quantitative Data.* 7th ed., Zeist, The Netherlands: TNO Biotechnology and Chemistry Institute.
2. Chang, S. S. 1989. Food flavors. *Food Technol* 43:99–106.
3. Emberger, R. 1985. An analytical approach to flavor research. *Cereal Foods World* 30:691–694.
4. Fischetti, F. 2005. Flavors. In *Kirk-Othmer Encyclopedia of Chemical Technology.* 5th ed., 11, ed. A. Seidel, 563–588. Hoboken, NJ: Wiley-Interscience.
5. Darriet, P., T. Tominaga, V. Lavigne, J.-N. Biodron, and D. Dubourdieu. 1995. Identification of a powerful aromatic component of *Vitis vinifera* L. var. Sauvignon wines: 4-mercapto-4-methylpentan-2-one, *Flav Frag J* 10:385–392.
6. McGorrin, R. J. 2007. Character impact flavor compounds: Flavors and off-flavors in foods. In *Sensory-Directed Flavor Analysis*, ed. R. Marsili, 223–267. Boca Raton, FL: Taylor & Francis.
7. Maarse, H. 1991. Introduction. In *Volatile Compounds in Foods and Beverages*, ed. H. Maarse, 1–40. New York, NY: Marcel Dekker.
8. Grosch, W. 1998. Flavor of coffee. A review. *Nahrung* 42:344–350.
9. Mayol, A. R., and T. E. Acree. 2001. Advances in gas chromatography-olfactometry. In *Gas Chromatography-Olfactometry: The State of the Art*, ed. J. V. Leland, P. Schieberle, A. Buettner, and T. E. Acree, 1–10. ACS Symposium Series 782. Washington, DC: American Chemical Society.
10. Mistry, B. H., T. Reineccius, and L. K. Olson. 1997. Gas chromatography-olfactometry for the determination of key odorants in foods. In *Techniques for Analyzing Food Aroma*, ed. R. Marsili, 265–292. New York: Marcel Dekker.
11. Pickenhagen, W. 1999. Flavor chemistry—the last 30 years. In *Flavor Chemistry: Thirty Years of Progress*, ed. R. Teranishi, E. L. Wick, and I. Hornstein, 75–87. New York, NY: Kluwer Academic/Plenum.
12. Boelens, M. H. 1991. Spices and condiments II. In *Volatile Compounds in Foods and Beverages*, ed. H. Maarse, 449–482. New York, NY: Marcel Dekker.
13. Diaz-Maroto, M., I. Hidalgo, E. Sanchez-Palomo, and M. Perez-Coello. 2005. Volatile components and key odorants of fennel (*Foeniculum vulgare* Mill.) and thyme (*Thymus vulgaris* L.) oil extracts obtained by simultaneous distillation-extraction and supercritical fluid extraction. *J Agric Food Chem* 53:5385–5389.
14. Bauer, K., D. Garbe, and H. Surburg. 2001. *Common Fragrance and Flavor Materials: Preparation, Properties, and Uses.* 4th ed. Weinheim, Germany: Wiley-VCH.
15. Blank, I., J. Lin, S. Devaud, R. Fumeaux, and L. B. Fay. 1997. The principal flavor components of fenugreek (*Trigonellla foenum-graecum* L.). In *Spices: Flavor Chemistry and Antioxidant Properties*, ed. S. J. Risch, and C.-T. Ho, 12–28. Washington, DC: ACS Symposium Series 660, American Chemical Society.
16. Hiserodt, R. D., C.-T. Ho, and R. T. Rosen. 1997. The characterization of volatile and semivolatile components in powdered turmeric by direct

thermal extraction gas chromatography—mass spectrometry. In *Spices: Flavor Chemistry and Antioxidant Properties*, ed. S. J. Risch, and C.-T. Ho, 80–97. Washington, DC: ACS Symposium Series 660, American Chemical Society.

17. Cadwallader, K. R., H. H. Baek, and M. Cai. 1997. Characterization of saffron flavor by aroma extract dilution analysis. In *Spices: Flavor Chemistry and Antioxidant Properties*, ed. S. J. Risch, and C.-T. Ho, 66–79. Washington, DC: ACS Symposium Series 660, American Chemical Society.

18. Rowe, D. 2000. More Fizz for your buck: High-impact aroma chemicals. *Perf Flav* 25:1–19.

19. Jagella, T., and W. Grosch. 1999. Flavor and off-flavor compounds of black and white pepper (*Piper nigrum* L.). II. Odor activity values of desirable and undesirable odorants of black pepper. *Eur Food Res Technol* 209:22–26.

20. Wood, C., T. E. Siebert, M. Parker, D. L. Capone, G. M. Elsey, A. P. Pollnitz, M. Eggers, M. Meier, T. Vossing, S. Widder, G. Krammer, M. A. Sefton, and M. J. Herderich. 2008. From wine to pepper: rotundone, an obscure sesquiterpene, is a potent spicy aroma compound. *J Agric Food Chem* 56:3738–3744.

21. Blank, I., A. Sen, and W. Grosch. 1992. Sensory study on the character-impact flavor compounds of dill herb (*Anethum graveolens* L.). *Food Chem* 43:337–343.

22. Boelens, M. H., and L. J. van Gemert. 1993. Volatile character-impact sulfur compounds and their sensory properties. *Perf Flav* 18:29–39.

23. Randle, W. M. 1997. Onion flavor chemistry and factors influencing flavor intensity. In *Spices: Flavor Chemistry and Antioxidant Properties*, ed. S. J. Risch, and C.-T. Ho, 41–52. Washington, DC: ACS Symposium Series 660, American Chemical Society.

24. Block, E. 2010. *Garlic and Other Alliums: The Lore and the Science*. Cambridge, UK: RSC Publishing.

25. Block, E., S. Naganathan, D. Putman, and S.-H. Zhao. 1992. Allium chemistry: HPLC analysis of thiosulfinates from onion, garlic, wild garlic (Ramsoms), leek, scallion, shallot, elephant (great-headed) garlic, chive, and Chinese chive. *J Agric Food Chem* 40:2418–2430.

26. Widder, S., C. S. Luntzel, T. Dittner, and W. Pickenhagen. 2000. 3-Mercapto-2-methylpentan-1-ol, a new powerful aroma compound. *J Agric Food Chem* 48:418–423.

27. Granvogl, M., M. Christlbauer, and P. Schieberle. 2004. Quantitation of the intense aroma compound 3-mercapto-2-methylpentan-1-ol in raw and processed onions (*Allium cepa*) of different origins and in other *Allium* varieties using a stable isotope dilution assay. *J Agric Food Chem* 52:2797–2802.

28. Majcher, M. A., and H. J. Jelen. 2005. Identification of potent odorants formed during the preparation of extruded potato snacks. *J Agric Food Chem* 53:6432–6437.

29. (a) Breme, K., N. Guillamon, X. Fernandez, P. Tournayre, H. Brevard, D. Joulain, J. L. Berdague, and U. J. Meierhenrich. 2009. First identification of O,S-diethyl thiocarbonate in Indian cress absolute and odor evaluation of its synthesized homologues by GC-sniffing. *J Agric Food Chem* 57:2503–2507; (b) Breme, K., N. Guillamon, X. Fernandez, P. Tournayre, H. Brevard, D. Joulain, J. L. Berdague, and U. J. Meierhenrich. 2010. Characterization of volatile compounds of Indian cress absolute by GC-olfactometry/

VIDEO-Sniff and comprehensive two-dimensional gas chromatography. *J Agric Food Chem* 58:473–480.

30. Berger, R. C. 1991. Fruits I. In *Volatile Compounds in Foods and Beverages*, ed. H. Maarse, 283–304. New York, NY: Marcel Dekker.

31. Roberts, D. D., and T. E. Acree. 1995. Developments in the isolation and characterization of β-damascenone precursors from apples. In *Fruit Flavors: Biogenesis, Characterization, and Authentication*, ed. R. L. Rouseff, and M. M. Leahy, 190–199. Washington, DC: ACS Symposium Series 596, American Chemical Society.

32. Larsen, M., L. Poll, and C. E. Olsen. 1992. Evaluation of the aroma composition of some strawberry (*Fragaria ananassa* Duch) cultivars by use of odor threshold values. *Z Lebensm Unters Forsch* 195:536–539.

33. Baek, H. H., and K. R. Cadwallader. 1999. Contribution of free and glycosidically bound volatile compounds to the aroma of muscadine grape juice. *J Food Sci* 64:441–444.

34. Schieberle, P., and T. Hofmann. 1997. Evaluation of the character impact odorants in fresh strawberry juice by quantitative measurements and sensory studies on model mixtures. *J Agric Food Chem* 45:227–232.

35. Kolor, M. G. 1983. Identification of an important new flavor compound in concord grapes. *J Agric Food Chem* 31:1125–1127.

36. Thorngate, J. H. 1998. Yeast strain and wine flavor: nature or nurture? In *Chemistry of Wine Flavor*, ed. A. L. Waterhouse, and S. E. Ebeler, 66–80. Washington, DC: ACS Symposium Series 714, American Chemical Society.

37. Fang, Y., and M. Qian. 2005. Aroma compounds in Oregon Pinot Noir wine determined by aroma extract dilution analysis (AEDA). *Flavour Fragr J* 20:22.

38. Campo, E., V. Ferreira, A. Escudero, and J. Cacho. 2005. Prediction of the wine sensory properties related to grape variety from dynamic-headspace gas chromatography—olfactometry data. *J Agric Food Chem* 53:5682–5690.

39. Mateo-Vivaracho, L., J. Zapata, J. Cacho, and V. Ferreira. 2010. Analysis, occurrence, and potential sensory significance of five polyfunctional mercaptans in white wines. *J Agric Food Chem* 58:10184–10194.

40. Farina, L., E. Boido, F. Carrau, G. Versini, and E. Dellacassa. 2005. Terpene compounds as possible precursors of 1,8-cineole in red grapes and wines. *J Agric Food Chem* 53:1633–1636.

41. (a) Guth, H. 1997. Identification of character impact odorants of different white wine varieties. *J Agric Food Chem* 45:3022–3026; (b) Guth, H. 1997. Quantitation and sensory studies of character impact odorants of different white wine varieties. *J Agric Food Chem* 45:3027–3032.

42. Kumazawa, K., K. Kubota, and H. Masuda. 2005. Influence of manufacturing conditions and crop season on the formation of 4-mercapto-4-methyl-2-pentanone in Japanese green tea (sen-cha). *J Agric Food Chem* 53:5390–5396.

43. Kishimoto, T., M. Kobayashi, N. Yako, A. Iida, and A. Wanikawa. 2008. Comparison of 4-mercapto-4-methylpentan-2-one contents in hop cultivars from different growing regions. *J Agric Food Chem* 56:1051–1057.

44. Buettner, A., and P. Schieberle. 2001. Evaluation of key compounds in hand-squeezed grapefruit juice (*Citrus paradisi* Macfayden) by quantitation and flavor reconstitution experiments. *J Agric Food Chem* 49:1358–1363.

45. Tressl, R., D. Bahri, and M. Kossa. 1980. Formation of off-flavor components in beer. In *The Analysis and Control of Less Desirable Flavors in Foods and Beverages*, ed. G. Charlambous, 293. New York, NY: Academic Press.

46. Boccorh, R. K., A. Paterson, and J. R. Piggot. 2002. Extraction of aroma components to quantify overall sensory character in a processed blackcurrant (*Ribes nigrum* L.) concentrate. *Flavour Fragr J* 17:385.
47. Goeke, A. 2002. Sulfur-containing odorants in fragrance chemistry. *Sulfur Reports* 23:243–278.
48. Demole, E., P. Enggist, and G. Ohloff. 1982. 1-*p*-Menthene-8-thiol: A powerful flavor impact constituent of grapefruit juice (*Citrus paradisi* MacFayden). *Helv Chim Act* 65:1785.
49. Lin, J., P. Jella, and R. L. Rouseff. 2002. Gas chromatography-olfactometry and chemiluminescence characterization of grapefruit juice volatile sulfur compounds. In *Heteroatomic Aroma Compounds*, ed. G. A. Reineccius, and T. A. Reineccius, 102–112. Washington, DC: ACS Symposium Series 826, American Chemical Society.
50. Mahattanatawee, K., R. Rouseff, M. F. Valim, and M. Naim. 2005. Identification and aroma impact of norisoprenoids in orange juice. *J Agric Food Chem* 53:393–397.
51. Shaw, P. E. 1991. Fruits II. In *Volatile Compounds in Foods and Beverages*, ed. H. Maarse, 305–328. New York, NY: Marcel Dekker.
52. Choi, H.-S. 2005. Characteristic odor components of kumquat (*Fortunella japonica* Swingle) peel oil. *J Agric Food Chem* 53:1642–1647.
53. Larsen, M., and L. Poll. 1990. Odor thresholds of some important aroma compounds in raspberries. *Z Lebensm Unters Forsch* 191:129–131.
54. Klesk, K., M. Qian, and R. R. Martin. 2004. Aroma extract dilution analysis of cv. Meeker (*Rubus idaeus* L.) red raspberries from Oregon and Washington. *J Agric Food Chem* 52:5155–5161.
55. Klesk, K., and M. Qian. 2003. Aroma extract dilution analysis of cv. Marion (*Rubus* spp. *hyb*) and cv. Evergreen (*R. laciniatus* L.) blackberries. *J Agric Food Chem* 51:3436–3441.
56. Wang, Y., C. Finn, and M. Qian. 2005. Impact of growing environment on Chickasaw blackberry (*Rubus* L.) aroma evaluated by gas chromatography olfactometry dilution analysis. *J Agric Food Chem* 53:3563–3571.
57. Engel, K. H., and R. Tressl. 1991. Identification of new sulfur-containing volatiles in yellow passion fruit (*Passiflora edulis f. flavicarpa*). *J Agric Food Chem* 39:2249–2252.
58. Werkhoff, P., M. Guntert, G. Krammer, H. Sommer, and J. Kaulen. 1998. Vacuum headspace method in aroma research: flavor chemistry of yellow passion fruits. *J Agric Food Chem* 46:1076–1093.
59. Sarrazin, E., S. Shinkaruk, M. Pons, C. Thibon, B. Bennetau, and P. Darriet. 2010. Elucidation of the 1,3-sulfanylalcohol oxidation mechanism: an unusual identification of the disulfide of 3-sulfanylhexanol in Sauternes botrytized wines. *J Agric Food Chem* 58:10606–10613.
60. (a) Steinhaus, M., D. Sinuco, J. Polster, C. Osorio, P. Schieberle. 2008. Characterization of the aroma-active compounds in pink guava (*Psidium guajava*, L.) by application of the aroma extract dilution analysis. *J Agric Food Chem* 56:4120–4127; (b) Steinhaus, M., D. Sinuco, J. Polster, C. Osorio, P. Schieberle. 2009. Characterization of the key aroma compounds in pink guava (*Psidium guajava* L.) by means of aroma re-engineering experiments and omission tests. *J Agric Food Chem* 57:2882–2888.
61. Weenen, H., Koolhaas, W. E., and A. Apriyantono. 1996. Sulfur-containing volatiles of durian fruits (*Durio zibethinus* Murr.). *J Agric Food Chem* 44:3291–3293.

62. Buttery, R. G. 1993. Quantitative and sensory aspects of flavor of tomato and other vegetables and fruits. In *Flavor Science: Sensible Principles and Techniques*, ed. T. E. Acree, and R. Teranishi, 259–286. Washington, DC: ACS Books, American Chemical Society.

63. Jordan, M. J., P. E. Shaw, and K. L. Goodner. 2001. Volatile components in aqueous essence and fresh fruit of *Cucumis melo* cv. Athena (muskmelon) by GC–MS and GC–O. *J Agric Food Chem* 49:5929–5933.

64. Guntert, M., G. Krammer, H. Sommer, and P. Werkhoff. 1998. The importance of the vacuum headspace method for the analysis of fruit flavors. In *Flavor Analysis: Developments in Isolation and Characterization*, ed. C. J. Mussinan, and M. J. Morello, 38–60. Washington, DC: ACS Symposium Series 705, American Chemical Society.

65. Whitfield, F. B., and J. H. Last. 1991. Vegetables. In *Volatile Compounds in Foods and Beverages*, ed. H. Maarse, 203–282. New York, NY: Marcel Dekker.

66. Takeoka, G. 1999. Flavor chemistry of vegetables. In *Flavor Chemistry: Thirty Years of Progress*, ed. R. Teranishi, E. L. Wick, and I. Hornstein, 287–304. New York, NY: Kluwer Academic/Plenum.

67. Naef, R., A. Velluz, and A. Jaquier. 2008. New volatile sulfur-containing constituents in a simultaneous distillation–extraction extract of red bell peppers (*Capsicum annuum*). *J Agric Food Chem* 56:517–527.

68. Schieberle, P., S. Ofner, and W. Grosch. 1990. Evaluation of potent odorants in cucumbers (*Cucumis sativus*) and muskmelons (*Cucumis melo*) by aroma extract dilution analysis. *J Food Sci* 55:193–195.

69. McGorrin, R. J., and L. Gimelfarb. 1998. Comparison of flavor components in fresh and cooked tomatillo with red plum tomato. In *Food Flavors: Formation, Analysis, and Packaging Influences*, ed. E. T. Contis, C.-T. Ho, C. J. Mussinan, T. H. Parliment, F. Shahidi, and A. M. Spanier, 295–313. New York, NY: Elsevier. Developments in Food Science 40.

70. Marks, H. S., J. A. Hilson, H. C. Leichtweis, and G. S. 1992. Stoewsand, S-Methylcysteine sulfoxide in *Brassica* vegetables and formation of methyl methanethiosulfinate from Brussels sprouts. *J Agric Food Chem* 40:2098–2101.

71. Chin, H.-W., and R. C. Lindsay. 1994. Mechanisms of formation of volatile sulfur compounds following the action of cysteine sulfoxide lyases. *J Agric Food Chem* 42:1529–1536.

72. Ulrich, D., E. Hoberg, T. Bittner, W. Engewald, and K. Meilchen. 2001. Contribution of volatile compounds to the flavor of cooked asparagus. *Eur Food Res Technol* 213:200–204.

73. Wagner, R. K., and W. Grosch. 1998. Key odorants of French fries. *J Amer Oil Chem Soc* 75:1385–1392.

74. Pelusio, F., T. Nilsson, L. Montanarella, R.Tilio, B. Larsen, S. Facchetti, and J. O. Madsen. 1995. Headspace solid-phase microextraction analysis of volatile organic sulfur compounds in black and white truffle aroma. *J Agric Food Chem* 43:2138–2143.

75. Bellesia, F., A. Pinetti, A. Bianchi, and B. Tirillini. 1996. Volatile compounds of the white truffle (*Tuber magnatum* Pico) from middle Italy. *Flav Frag J* 11:239–243.

76. Masanetz, C., and W. Grosch. 1998. Key odorants of parsley leaves (*Petroselinum crispum* [Mill.] Nym. *ssp. crispum*) by odor-activity values. *Flavor Fragr J* 13:115–124.

77. Scarpellino, R., and R. J. Soukup. 1993. Key flavors from heat reactions of food ingredients. In *Flavor Science: Sensible Principles and Techniques*, ed. T. E. Acree, and R. Teranishi, 309–335. Washington, DC: ACS Books, American Chemical Society.

78. Zabetakis, I., J. W. Gramshaw, and D. S. Robinson. 1999. 2,5-Dimethyl-4-hydroxy-2*H*-furan-3-one and its derivatives: analysis, synthesis and biosynthesis—a review. *Food Chem* 65:139–151.

79. Kobayashi, A. 1989. Sotolon: identification, formation, and effect on flavor. In *Flavor Chemistry: Trends and Developments*, ed. R. Teranishi, R. G. Buttery, and F. Shahidi, 49–59. Washington, DC: ACS Symposium Series 388, American Chemical Society.

80. Grosch, W., M. Czerny, F. Mayer, and A. Moors. 2000. Sensory studies on the key odorants of roasted coffee. In *Caffeinated Beverages: Health Benefits, Physiological Effects, and Chemistry*, ed. T. H. Parliment, C.-T. Ho, P. Schieberle, 202–209. Washington, DC: ACS Symposium Series 754, American Chemical Society.

81. Flament, I. 1991. Coffee, cocoa and tea. In *Volatile Compounds in Foods and Beverages*, ed. H. Maarse, 617–670. New York: Marcel Dekker.

82. Czerny, M., F. Mayer, and W. Grosch. 1999. Sensory study on the character impact odorants of roasted Arabica coffee. *J Agric Food Chem* 47:695–699.

83. Tominaga, T., and D. Dubourdieu. 2006. A novel method for quantification of 2-methyl-3-furanthiol and 2-furanmethanethiol in wines made from *Vitis vinifera* grape varieties. *J Agric Food Chem* 54:29–33.

84. Parliment, T. H., M. J. Morello, and R. J. McGorrin, ed. 1994. Washington, DC: ACS Symposium Series 543, American Chemical Society.

85. Schieberle, P. 1995. Quantitation of important roast-smelling odorants in popcorn by stable isotope dilution assays and model studies on flavor formation during popping. *J Agric Food Chem* 43:2442–2448.

86. Hofmann, T., and P. Schieberle. 1998. 2-Oxopropanal, hydroxyl-2-propanone, and 1-pyrroline—important intermediates in the generation of the roast-smelling food flavor compounds 2-acetyl-1-pyrroline and 2-acetyltetrahydropyridine. *J Agric Food Chem* 46:2270–2277.

87. Schieberle, P., and W. Grosch. 1989. Bread flavor. In *Thermal Generation of Aromas*, ed. T. H. Parliament, R. J. McGorrin, and C.-T. Ho, 258–267. Washington, DC: ACS Symposium Series 409, American Chemical Society.

88. Grosch, W., and P. Schieberle. 1997. Flavor of cereal products—a review. *Cereal Chem* 74:91–97.

89. Hofmann, T., R. Hassner, and P. Schieberle. 1995. Determination of the chemical structure of the intense roasty, popcorn-like odorant 5-acetyl-2,3-dihydro-1,4-thiazine. *J Agric Food Chem* 43:2195–2198.

90. Engel, W., and P. Schieberle. 2002. Structural determination and odor characterization of N-(2-mercaptoethyl)-1,3-thiazolidine, a new intense popcorn-like-smelling odorant. *J Agric Food Chem* 50:5391–5393.

91. Buttery, R. G., and L. C. Ling. 1994. Importance of 2-aminoacetophenone to the flavor of masa corn flour products. *J Agric Food Chem* 42:1–2.

92. Buttery, R. G., and L. C. Ling. 1998. Additional studies on flavor components of corn tortilla chips. *J Agric Food Chem* 46:2764–2769.

93. Heydanek, M. G., and R. J. McGorrin. 1981. Gas chromatography–mass spectrometry investigations on the flavor chemistry of oat groats. *J Agric Food Chem* 29:950–954.

94. Heydanek, M. G., and R. J. McGorrin. 1986. Oat flavor chemistry: principles and prospects. In *Oats: Chemistry and Technology*, ed. F. H. Webster, 335–369. St. Paul, MN: American Association of Cereal Chemists.
95. Schuh, C., and P. Schieberle. 2005. Characterization of (E,E,Z)-2,4,6-nonatrienal as a character impact aroma compound of oat flakes. *J Agric Food Chem* 53:8699–8705.
96. Zehentbauer, G., and W. Grosch. 1998. Crust aroma of baguettes. I. Key odorants of baguettes prepared in two different ways. *J Cereal Sci* 28:81–92.
97. Hofmann, T., and P. Schieberle. 1998. Identification of key aroma compounds generated from cysteine and carbohydrates under roasting conditions. *Z Lebensm Unters Forsch* 207:229–236.
98. Schnermann, P., and P. Schieberle. 1997. Evaluation of key odorants in milk chocolate and cocoa mass by aroma extract dilution analyses. *J Agric Food Chem* 45:867–872.
99. van Praag, M., H. S. Stein, and M. S. 1968. Tibetts. Steam volatile constituents of roasted cocoa beans. *J Agric Food Chem* 16:1005–1008.
100. Burdock, G. A., ed. 2005. *Fenaroli's Handbook of Flavor Ingredients*, 5th ed., 2. CRC Press: Boca Raton, FL.
101. Frauendorfer, F., and P. Schieberle. 2006. Identification of the key aroma compounds in cocoa powder based on molecular sensory correlations. *J Agric Food Chem* 54:5521–5529.
102. Maga, J. 1991. Nuts. In *Volatile Compounds in Foods and Beverages*, ed. H. Maarse, 671–688. New York, NY: Marcel Dekker.
103. Burdack-Freitag, A., and P. Schieberle. 2010. Changes in the key odorants of Italian hazelnuts (*Coryllus avellana* L. var. Tonda Romana) induced by roasting. *J Agric Food Chem* 58:6351–6359.
104. Withycombe, D. A., and C. J. Mussinan. 1988. Identification of 2-methyl-3-furanthiol in the steam distillate from canned tuna fish. *J Food Sci* 53: 658–659.
105. Mottram, D. S. 1991. Meat. In *Volatile Compounds in Foods and Beverages*, ed. H. Maarse, 107–178. New York, NY: Marcel Dekker.
106. Cerny, C., and W. Grosch. 1993. Quantification of character-impact odor compounds of roasted beef. *Z Lebensm Unters Forsch* 196:417–422.
107. Cerny, C., and W. Grosch. 1994. Precursors of ethyldimethylpyrazine isomers and 2,3-diethyl-5-methylpyrazine formed in roasted beef. *Z Lebensm Unters Forsch* 198:210–214.
108. Kerler, J., and W. Grosch. 1997. Character impact odorants of boiled chicken: changes during refrigerated storage and reheating. *Z Lebensm Unters Forsch* 205:232–238.
109. Guth, H., and W. Grosch. 1993. 12-Methyltridecanal, a species-specific odorant of stewed beef. *Lebensm Wiss und Technol* 26:171–177.
110. Kerler, J., and W. Grosch. 1996. Odorants contributing to warmed-over flavor (WOF) of refrigerated cooked beef. *J Food Sci* 61:1271–1275.
111. Josephson, D. B. 1991. Seafood. In *Volatile Compounds in Foods and Beverages*, ed. H. Maarse, 179–202. New York, NY: Marcel Dekker.
112. Shahidi, F., and K. R. Cadwallader. ed. 1997. *Flavor and Lipid Chemistry of Seafoods*. Washington, DC: ACS Symposium Series 674, American Chemical Society.
113. Gordon, R. J., and D. B. Josephson. 2000. Surimi seafood flavors. In *Surimi and Surimi Seafood*, ed. J. W. Park, 393–416. New York, NY: Marcel Dekker.

114. Milo, C., and W. Grosch. 1996. Changes in the odorants of boiled salmon and cod as affected by the storage of the raw material. *J Agric Food Chem* 44:2366–2371.

115. Milo, C., and W. Grosch. 1993. Changes in the odorants of boiled trout (*Salmo fario*) as affected by the storage of the raw material. *J Agric Food Chem* 41: 2076–2081.

116. (a) Karahadian, C., and R. C. Lindsay. 1989. Evaluation of compounds characterizing fishy flavors in fish oils. *J Amer Oil Chem Soc* 66:953–960; (b) Kobayashi, A., K. Kubota, M. Iwamoto, and H. Tamura. 1989. Syntheses and sensory characterization of 5,8,11-tetradeca-trien-2-one isomers. *J Agric Food Chem* 37:151–154.

117. Sekiwa, Y., K. Kubota, and A. Kobayashi. 1997. Characteristic flavor components in the brew of cooked clam (*Meretrix lusoria*) and the effect of storage on flavor formation. *J Agric Food Chem* 45:826–830.

118. Baek, H. H., and K. R. Cadwallader. 1997. Character-impact aroma compounds of crustaceans. In *Flavor and Lipid Chemistry of Seafoods*, ed. F. Shahidi, and K. R. Cadwallader, 85–94. Washington, DC: ACS Symposium Series 674, American Chemical Society.

119. Lee, G.-H., O Suriyaphan, and K. R. Cadwallader. 2001. Aroma components of cooked tail meat of American lobster (*Homarus americanus*). *J Agric Food Chem* 49:4324–4332.

120. Kubota, K., A. Nakamoto, M. Moriguchi, A. Kobayashi, and H. Ishii. 1991. Formation of Pyrrolidino[1,2-*e*]-4*H*-2,4-dimethyl-1,3,5-dithiazine in the volatiles of boiled short-necked clam, clam, and corbicula. *J Agric Food Chem* 39:1127–1130.

121. Whitfield, F. B., M. Drew, F. Helidoniotis, and D. Svoronos. 1999. Distribution of bromophenols in species of marine polychaetes and bryozoans from eastern Australia and the role of such animals in the flavor of edible ocean fish and prawns (shrimp). *J Agric Food Chem* 47:4756–4762.

122. Schutte, L. 1999. Development and application of dairy flavors. In *Flavor Chemistry: Thirty Years of Progress*, ed. R. Teranishi, E. L. Wick, and I. Hornstein, 155–165. New York, NY: Kluwer Academic/Plenum.

123. Parliment, T. H., and R. J. McGorrin. 2000. Critical flavor compounds in dairy products. In *Flavor Chemistry: Industrial and Academic Research*, ed. S. J. Risch, C.-T. Ho, 44–71. Washington, DC: ACS Symposium Series 756, American Chemical Society.

124. McGorrin, R. J. 2001. Advances in dairy flavor chemistry. In *Food Flavors and Chemistry: Advances of the New Millennium*, ed. A. M. Spanier, F. Shahidi, T. H. Parliment, C. J. Mussinan, C.-T. Ho, and E. T. Contis, 67–84. Cambridge: Royal Society of Chemistry.

125. Marsili, R. 2003. Flavors and off-flavors in dairy foods. In *Encyclopedia of Dairy Sciences*, ed. H. Roginski, J. Fuquay, and P. F. Fox, 1069–1081. New York, NY: Elsevier.

126. Singh, T. K., M. A. Drake, and K. R. Cadwallader. 2003. Flavor of cheddar cheese: a chemical and sensory perspective. *Compr Rev Food Sci Food Saf* 2:139–162.

127. Urbach, G. 1997. The flavor of milk and dairy products: II. Cheese: contribution of volatile compounds. *Int J Dairy Technol* 50:79–89.

128. Sable, S., and G. Cottenceau. 1999. Current knowledge of soft cheeses flavor and related compounds. *J Agric Food Chem* 47:4825–4836.

129. Sourabie, A. M., H.-E. Spinnler, P. Bonnarme, A. Saint-Eve, and S. Landaud. 2008. Identification of a powerful aroma compound in Munster and Camembert cheeses: ethyl 3-mercaptopropionate. *J Agric Food Chem* 56: 4674–4680.
130. Carunchia Whetstine, M. E., Y. Karagul-Yuceer, Y. K. Avsar, and M. A. Drake. 2003. Identification and quantification of character aroma components in fresh chevre-style goat cheese. *J Food Sci* 68:2441–2447.
131. Qian, M., and G. A. Reineccius. 2003. Quantification of aroma compounds in Parmigiano Reggiano cheese by a dynamic headspace gas chromatography— mass spectrometry technique and calculation of odor activity value. *J Dairy Sci* 86:770–776.
132. Qian, M., and G. Reineccius. 2003. Static headspace and aroma extract dilution analysis of Parmigiano Reggiano cheese. *J Food Sci* 68:794–798.
133. Qian, M., and G. Reineccius. 2003. Potent aroma compounds in Parmigiano Reggiano cheese studied using a dynamic headspace (purge-trap) method. *Flavour Fragr J* 18:252–259.
134. Milo, C., and G. A. Reineccius. 1997. Identification and quantification of potent odorants in regular-fat and low-fat mild cheddar cheese. *J Agric Food Chem* 45:3590–3594.
135. Zehentbauer, G., and G. A. Reineccius. 2002. Determination of key aroma components of cheddar cheese using dynamic headspace dilution assay. *Flavour Fragr J* 17:300–305.
136. Burbank, H. M., and M. C. Qian. 2005. Volatile sulfur compounds in Cheddar cheese determined by headspace solid-phase microextraction and gas chromatograph-pulsed flame photometric detection. *J Chromatogr A* 1066:149–157.
137. Burbank, H., and M. C. Qian. 2008. Development of volatile sulfur compounds in heat-shocked and pasteurized milk cheese. *Int Dairy J* 18:811–818.
138. Avsar, Y. K., Y. Karagul-Yuceer, M. A. Drake, T. K. Singh, Y. Yoon, and K. R. Cadwallader. 2004. Characterization of nutty flavor in cheddar cheese. *J Dairy Sci* 87:1999–2010.
139. Kleinhenz, J. K., C. J. Kuo, and W. J. Harper. 2007. Evaluation of polyfunctional thiol compounds in aged Cheddar cheese: estimated concentrations. *Milchwissenschaft* 62:181–183.
140. Marsili, R. 2007. Application of solid-phase microextraction gas chromatography-mass spectrometry for flavor analysis of cheese-based products. In *Flavor of Dairy Products*, ed. K. Cadwallader, M. A. Drake, and R. J. McGorrin, 79–91. Washington, DC: ACS Symposium Series 971, American Chemical Society.
141. Schieberle, P., K. Gassenmeier, H. Guth, A. Sen, and W. Grosch. 1993. Character impact odor compounds of different kinds of butter. *Lebensm Wiss und Technol* 26:347–356.
142. Karagul-Yuceer, Y., M. A. Drake, and K. R. Cadwallader. 2001. Aroma-active components of nonfat dry milk. *J Agric Food Chem* 49:2948–2953.
143. Mahajan, S. S., L. Goddik, and M. C. Qian. 2004. Aroma compounds in sweet whey powder. *J Dairy Sci* 87:4057–4063.
144. Carunchia Whetstine, M. E., A. E. Croissant, and M. A. Drake. 2005. Characterization of dried whey protein concentrate and isolate flavor. *J Dairy Sci* 88:3826–3839.

145. Budin, J. T., C. Milo, and G. A. Reineccius. 2001. Perceivable odorants in fresh and heated sweet cream butters. In *Food Flavors and Chemistry: Advances of the New Millennium*, ed. A. M. Spanier, F. Shahidi, T. H. Parliment, C. J. Mussinan, C.-T. Ho, and E. T. Contis, 85–96. Cambridge: Royal Society of Chemistry.

146. (a) Peterson, D. G., and G. A. Reineccius. 2003. Characterization of the volatile components which constitute fresh sweet cream butter aroma. *Flavour Frag J* 18:215–220; (b) Peterson, D. G., and G. A. Reineccius. 2003. Determination of the aroma impact compounds in sweet cream butter. *Flavour Frag J* 18:320–324.

147. Preininger, M., and W. Grosch. 1994. Evaluation of key odorants of the neutral volatiles of Emmentaler cheese by the calculation of odor activity values. *Lebensm Wiss und Technol* 27:237–244.

148. Ott, A., L. B. Fay, and A. Chaintreau. 1997. Determination and origin of the aroma impact compounds of yogurt flavor. *J Agric Food Chem* 45:850–858.

149. Saxby, M. J. 1996. *Food Taints and Off-Flavors*, 2nd ed. Blackie Acad. Prof.: London.

150. Reineccius, G. 2006. Off-flavors and taints in foods. In *Flavor Chemistry and Technology*, 2nd ed. 161–200. Boca Raton, FL: CRC Press/Taylor & Francis.

151. Nijssen, B. 1991. Off-flavors. In *Volatile Compounds in Food and Beverages*, ed. H. Maarse, 689–736. New York: Marcel Dekker.

152. Charalambous, G. ed. 1992. *Off-Flavors in Foods and Beverages*. New York: Elsevier.

153. Marsili, R. 1997. Off-flavors and malodors in foods: Mechanisms of formation and analytical techniques. In *Techniques for Analyzing Food Aroma*, ed. R. Marsili, 237–264. New York, NY: Marcel Dekker.

154. Widder, S., and W. Grosch. 1997. Precursors of 2-nonenals causing the cardboard off-flavor in butter oil. *Nahrung* 41:42–45.

155. (a) Stephan, A., and H. Steinhart. 1999. Identification of character impact odorants of different soybean lecithins. *J Agric Food Chem* 47:2854–2859; (b) Boatright, W. L., and Q. Lei. 1999. Compounds contributing to the beany odor of aqueous solutions of soy protein isolates. *J Food Sci* 64: 667–670.

156. (a) Guth, H., and W. Grosch. 1989. 3-Methylnonane-2,4-dione—An intense odor compound formed during flavor reversion of soybean oil. *Lipid/Fett* 91:225–230; (b) Guth, H., and W. Grosch. 1999. Impact of 3-methylnonane-2,4-dione on the flavor of oxidized soybean oil. *J Amer Oil Chem Soc* 76:145.

157. Neff, W. E., and E. Selke. 1993. Volatile compounds from the triacylglycerol of *cis,cis* 9,15-linoleic acid. *J Amer Oil Chem Soc* 70:157–161.

158. Czerny, M., and A. Buettner. 2009. Odor-active compounds in cardboard. *J Agric Food Chem* 57:9979–9984.

159. Noel, S., C. Liegeois, G. Lermusieau, E. Bodart, C. Badot, and S. Collin. 1999. Release of deuterated nonenal during beer aging from labeled precursors synthesized in the boiling kettle. *J Agric Food Chem* 47:4323–4326.

160. Harayama, K., F. Hayase, and H. Kato. 1995. Contribution to stale flavor of 2-furfuryl ethyl ether and its formation mechanism in beer. *Biosci Biotech Biochem* 59:1144–1146.

161. Callemien, D., S. Dasnoy, and S. Collin. 2006. Identification of a stale-beer-like odorant in extracts of naturally aged beer. *J Agric Food Chem* 54:1409–1413.

162. Coghe, S., K. Benoot, F. Delvaux, B. Vanderhaegen, and F. R. Delvaux. 2004. Ferulic acid release and 4-vinylguaiacol formation during brewing and fermentation: indications for feruloyl esterase activity in *Saccharomyces cerevisiae. J Agric Food Chem* 52:602–608.
163. Jensen, K., M. A. Petersen, L. Poll, and P. B. Brockhoff. 1999. Influence of variety and growing location on the development of off-flavor in precooked vacuum-packed potatoes. *J Agric Food Chem* 47:1145–1149.
164. Masanetz, C., H. Guth, and W. Grosch. 1998. Fishy and hay-like off-flavors of dry spinach. *Z Lebensm Unters Forsch* 206:108–113.
165. Masanetz, C., H. Guth, and W. Grosch. 1998. Hay-like off-flavor of dry parsley. *Z Lebensm Unters Forsch* 206:114–120.
166. Kamimura, M., and H. Kaneda. 1992. Off-flavors in beer. In *Off-Flavors in Foods and Beverages*, ed. G. Charalambous, 433–472. New York: Elsevier.
167. Masuda, S., K. Kikuchi, K. Harayama, K. Sakai, and M. Ikeda. 2000. Determination of lightstruck character in beer by gas chromatography-mass spectrometry. *J Amer Soc Brew Chem* 58:152–154.
168. Templar, J., K. Arrigan, and W. J. Simpson. 1995. Formation, measurement and significance of light-struck flavor in beer: a review. *Brewers Digest* 70:18–25.
169. Lermusieau, G., and S. Collin. 2003. Volatile sulfur compounds in hops and residual concentrations in beer—a review. *J Am Soc Brew Chem* 61:109–113.
170. Clapperton, J. F. 1976. Ribes flavor in beer. *J Inst Brew* 82:175–176.
171. Schieberle, P. 1991. Primary odorants of pale lager beer: differences to other beers and changes during storage. *Z Lebensm UntersForsch* 193:558–565.
172. Escudero, D. A., P. Hernandez-Orte, J. Cacho, and V. Ferreira. 2000. Clues about the role of methional as character impact odorant of some oxidized wines. *J Agric Food Chem* 48:4268–4272.
173. Cullere, L., J. Cacho, and V. Ferreira. 2007. An assessment of the role played by some oxidation-related aldehydes in wine aroma. *J Agric Food Chem* 55:876–881.
174. Bezman, Y., R. L. Rouseff, and M. Naim. 2001. 2-Methyl-3-furanthiol and methional are possible off-flavors in stored orange juice: aroma-similarity, NIF/SNIF GC–O, and GC analyses. J Agric Food Chem 49:5425–5432.
175. Perez-Cacho, P. R., K. Mahattanatawee, J. M. Smoot, and R. Rouseff. 2007. Identification of sulfur volatiles in canned orange juices lacking orange flavor. *J Agric Food Chem* 55:5761–5767.
176. Perpete, P., and S. Collin. 1999. Contribution of 3-methylthiopropionaldehyde to the worty flavor of alcohol-free beers. *J Agric Food Chem* 47:2374–2378.
177. Lozano, P. R., M. A. Drake, D. Benitez, and K. R. Cadwallader. 2007. Instrumental and sensory characterization of heat-induced odorants in aseptically packaged soy milk. *J Agric Food Chem* 55:3018–3026.
178. Colahan-Sederstrom, P., and G. D. Peterson. 2005. Inhibition of key aroma compound generated during ultrahigh-temperature processing of bovine milk via epicatechin addition. *J Agric Food Chem* 53:398–402.
179. Jung, M. Y., S. H. Yoon, H. O. Lee, and D. B. Min. 1998. Singlet oxygen and ascorbic acid effects on dimethyl disulfide and off-flavor in skim milk exposed to light. *J Food Sci* 63:408–412.
180. Swaine, R. L., T. S. Myers, C. M. Fischer, N. Meyer, and B. Pohlkamp. 1997. The formation of off-flavors in fortified juice-containing beverages: identification of 4-methyl-2-isopropylthiazole. *Perf Flav* 22:57–58, 60, 62, 64.
181. Bernath, K. 1998. Sulfurous off-flavours in wine. [Das Boeckser-aroma in Wein]. *Diss Abstr Intl C* 59:308.

182. Anocibar-Beloqui, A., P. Guedes de Pinho, and A. Bertrand. 1995. Bis (2-hydroxyethyl) disulfide, a new sulfur compound found in wine. Its influence in wine aroma. *Amer J Enol Viticult* 46:84–87.

183. Licker, J. L., T. E. Acree, and T. Henick-Kling. 1998. What is "Brett" (*Brettanomyces*) flavor?: A preliminary investigation. In *Chemistry of Wine Flavor*, ed. A. L. Waterhouse, and S. E. Ebeler, 96–115. Washington, DC: ACS Symposium Series 714, American Chemical Society.

184. Lindinger, C., P. Pollien, R. C. H. de Vos, Y. Tikunov, J. A. Hageman, C. Lambot, R. Fumeaux, E. Voirol-Baliguet, and I. Blank. 2009. Identification of ethyl formate as a quality marker of the fermented off-note in coffee by a nontargeted chemometric approach. *J Agric Food Chem* 57:9972–9978.

185. Chambers IV, E., E. C. Smith, L. M. Seitz, and D. B. Sauer. 1998. Sensory properties of musty compounds in food. In *Food Flavors: Formation, Analysis, and Packaging Influences*, ed. E. T. Contis, C.-T. Ho, C. J. Mussinan, T. H. Parliment, F. Shahidi, and A. M. Spanier, 173–180, New York, NY: Elsevier.

186. Whitfield, F. B., J. L. Hill, and K. J. Shaw. 1997. 2,4,6-Tribromoanisole: a potential cause of mustiness in packaged food. *J Agric Food Chem* 45: 889–893.

187. Jonsson, S., J. Hagberg, and B. Van Bavel. 2008. Determination of 2,4,6-trichloroanisole and 2,4,6-tribromoanisole in wine using microextraction in packed syringe and gas chromatography-mass spectrometry. *J Agric Food Chem* 56:4962–4967.

188. Aung, L. H., J. L. Smilanick, P. V. Vail, P. L. Hartsell, and E. Gomez. 1996. Investigations into the origin of chloroanisoles causing musty off-flavor of raisins. *J Agric Food Chem* 44:3294–3296.

189. Andrewes, P., J. G. Bendall, C. Davey, and R. Shingleton. 2010. A musty flavor defect in calcium caseinate due to chemical tainting by 2,4,6-tribromophenol and 2,4,6-tribromoanisole. *Int Dairy J* 20:423–428.

190. Spadone, J. C., G. Takeoka, and R. Liardon. 1990. Analytical investigation of Rio off-flavor in green coffee. *J Agric Food Chem* 38:226–233.

191. Vitzthum, O. G. 1999. Thirty years of coffee chemistry research. In *Flavor Chemistry: Thirty Years of Progress*, ed. R. Teranishi, E. L. Wick, I. Hornstein, 117–133. New York, NY: Kluwer Academic/Plenum.

192. Lloyd, S. W., and C. C. Grimm. 1999. Analysis of 2-methylisoborneol and geosmin in catfish by microwave distillation—solid phase microextraction. *J Agric Food Chem* 47:164–169.

193. Jirawan, Y., and N. Athapol. 2000. Geosmin and off-flavor in Nile tilapia (*Oreochromis niloticus*). *J Aquat Food Prod Technol* 9:29–41.

194. Sevenants, M. R., and R. A. Sanders. 1984. Anatomy of an off-flavor investigation: The "medicinal" cake mix. *Analytical Chem* 56:293A–298A.

195. McGorrin, R. J., T. R. Pofahl, and W. R. Croasmun. 1987. Identification of the musty component from an off-odor packaging film. *Analytical Chem* 59: 1109A–1112A.

196. Sanders, R. A., D. V. Zyzak, T. R. Morsch, S. P. Zimmerman, P. M. Searles, M. A. Strothers, B. L. Eberhart, and A. K. Woo. 2005. Identification of 8-nonenal as an important contributor to "plastic" off-odor in polyethylene packaging. *J Agric Food Chem* 53:1713–1716.

197. Parks, O. W., D. P. Schwartz, and M. Keeney. 1964. Identification of o-aminoacetophenone as a flavor compound in stale dry milk. *Nature* 202:185–187.

198. Ramshaw, E. H., and E. A. Dunstone. 1969. Volatile compounds associated with the off-flavor in stored casein. *J Dairy Res* 36:215–223.
199. Hoenicke, K., O. Borchert, K. Gruning, and T. J. Simat. 2002. "Untypical aging off-flavor" in wine: Synthesis of potential degradation compounds of indole-3-acetic acid and kynurenine and their evaluation as precursors of 2-aminoacetophenone. *J Agric Food Chem* 50:4303–4309.
200. Stark, W., and D. A. Forss. 1962. A compound responsible for metallic flavor in dairy products. I. Isolation and identification. *J Dairy Res* 29:173–180.
201. Lubran, M. B., H. T. Lawless, E. Lavin, and T. E. Acree. 2005. Identification of metallic-smelling 1-octen-3-one and 1-nonen-3-one from solutions of ferrous sulfate. *J Agric Food Chem* 53:8325–8327.
202. Koenig, T., B. Gutsche, M. Hartl, R. Huebscher, P. Schreier, and W. Schwab. 1999. 3-Hydroxy-4,5-dimethyl-2(5H)-furanone (sotolon) causing an off-flavor: elucidation of its formation pathways during storage of citrus soft drinks. *J Agric Food Chem* 47:3288–3291.
203. Buttery, R. G., W. F. Haddon, R. M. Seifert, and J. G. Turnbaugh. 1984. Thiamin odor and bis(2-methyl-3-furyl)disulfide. *J Agric Food Chem* 32:674–676.
204. Scarlata, C. J., and S. E. Ebeler. 1999. Headspace solid phase microextraction for the analysis of dimethyl sulfide in beer. *J Agric Food Chem* 47:2505–2508.
205. Heiler, C., and P. Schieberle. 1997. Model studies on the precursors and formation of the metallic smelling (E,Z)-2,6-nonadienol during the manufacture and storage of buttermilk. *Int Dairy J* 7:667–674.
206. Sensidoni, A., G. Rondinini, D. Peressini, M. Maifreni, and R. Bortolomeazzi. 1994. Presence of an off-flavor associated with the use of sorbates in cheese and margarine. *Ital J Food Sci* 6:237–242.
207. Shiratsuchi, H., Y. Yoshimura, M. Shimoda, K. Noda, and Y. Osajima. 1996. Objective evaluation of off-flavor in spray-dried skim milk powder. *J Jap Soc Food Sci Technol* 43:7–11.
208. Preininger, M., and F. Ullrich. 2001. Trace compound analysis for off-flavor characterization of micromilled milk powder. In *Gas Chromatography-Olfactometry: The State of the Art*, ed. J. V. Leland, P. Schieberle, A. Buettner, and T. E. Acree, 46–61. Washington, DC: ACS Symposium Series 782, American Chemical Society.
209. Karagul-Yuceer, Y., K. N. Vlahovich, M.A. Drake, and K. R. Cadwallader. 2003. Characteristic aroma components of rennet casein. *J Agric Food Chem* 51:6797–6801.
210. Iwatsuki, K., Y. Mizota, T. Kubota, O. Nishimura, H. Matsuda, K. Sotoyama, and M. Tomita. 1999. Evaluation of aroma of pasteurized and UHT processed milk by aroma extract dilution analysis. *J Jpn Soc Food Sci* 46:587–597.
211. Vazquez-Landaverde, P. A., G. Velazquez, J. A. Torres, and M. C. Qian. 2005. Quantitative determination of thermally derived off-flavor compounds in milk using solid-phase microextraction and gas chromatography. *J Dairy Sci* 88:3764–3772.
212. Whitfield, F. B., N. Jensen, and K. J. Shaw. 2000. Role of *Yersinia intermedia* and *Pseudomonas putida* in the development of a fruity off-flavor in pasteurized milk. *J Dairy Res* 67:561–569.
213. Zabolotsky, D. A., L. F. Chen, J. A. Patterson, J. C. Forrest, H. M. Lin, and A. L. Grant. 1995. Supercritical carbon dioxide extraction of androstenone and skatole from pork fat. *J Food Sci* 60:1006–1008.

214. Tanaka, Y., T. Sasao, T. Kirigaya, S. Hosoi, A. Mizuno, T. Kawamura, and H. Nakazawa. 1998. Analysis of skatole in off-flavor beef by GC/MS. *J Food Hygienic Soc Jap* 39:281–285.
215. Popken, A. M., H. M. Dechent, and D. Guerster. 1999. Investigations on the origin of carvone in orange juices as an off-flavor component. *Fruit Processing* 9:338–341.
216. Walsh, M., R. Rouseff, and M. Naim. 1997. Determination of furaneol and *p*-vinylguaiacol in orange juice employing differential UV wavelength and fluorescence detection with a unified solid phase extraction. *J Agric Food Chem* 45:1320–1324.
217. Donaghy, J. A., P. F. Kelly, and A. McKay. 1999. Conversion of ferulic acid to 4-vinyl guaiacol by yeasts isolated from unpasteurised apple juice. *J Sci Food Agric* 79:453–456.

Index